T0330159

SCIENCE AND POWER IN COLONIAL MAURITIUS

Rochester Studies in African History and the Diaspora

SCIENCE AND POWER
IN
COLONIAL MAURITIUS

William Kelleher Storey

UNIVERSITY OF ROCHESTER PRESS

First published 1997
Transferred to digital printing 2007

University of Rochester Press
668 Mt. Hope Avenue, Rochester, NY 14620, USA
www.urpress.com
and Boydell & Brewer Limited
PO Box 9, Woodbridge, Suffolk IP12 3DF, UK
www.boydellandbrewer.com

ISBN-13: 978-1-58046-015-6
ISBN-10: 1-58046-015-1
ISSN: 1092-5228

Library of Congress Cataloging-in-Publication Data

Storey, William K.
    Science and power in colonial Mauritius / William Kelleher Storey,
        p.   cm. — (Rochester studies in African history and the
    diaspora, ISSN 1092-5228 ; v. 3)
    Includes bibliographical references and index.
    ISBN 1-58046-015-1 (alk. paper)
    1. Sugarcane—Research—Mauritius—History. 2. Sugarcane—
    Economic aspects—Mauritius—History. 3. Sugarcane—Political
    aspects—Mauritius—History. 4. Mauritius—Economic conditions.
    5. Mauritius—Politics and government. I. Title. II. Series.
    SB229.M3S76   1997
    338.1'62—dc21   97-37387
                                                            CIP

A catalogue record for this title is available from the British Library.

Designed and typeset by Cornerstone Composition Services
This publication is printed on acid-free paper
Printed in the United States of America

CONTENTS

TABLES

PREFACE

This is a book about the history of science in Mauritius, a remote, densely populated island in the southwest Indian Ocean. Between 1815 and 1968, it became Britain's most important sugar-producing colony. Mauritians came to depend on sugar, and as a consequence they also came to depend on the scientists who supported sugar cane cultivation. I could have written about other sciences in Mauritius, such as astronomy, medicine, and meteorology, but sugar cane science was more important than any of these. Sugar cane scientists upheld the colonial economy, and they were also important for the political development of the island. Scientists and farmers from many different classes and cultures debated the production and distribution of new sugar cane plants. Their debates indicate the close relationship between science and politics in a colony.

Colleagues in England, Mauritius, and the United States helped me to write this book. The project began as my doctoral dissertation at Johns Hopkins, under the generous supervision of Philip Curtin. As the project evolved from a dissertation into a book, many other people commented on parts of five successive drafts. Some of them will continue to disagree with me on a number of points, but I still wish to express my gratitude. My thanks go to Samer Alatout, Rich Allen, Sara Berry, Jocelyn Chan Low, Arthur Daemmrich, Michael Dennis, Sam Gammon, Sheila Jasanoff, Franklin Knight, Jacques Koenig, Norma Kriger, Bill Leslie, Paul Paskoff, Archana Prasad, Sada Reddi, Jean-Claude Tyack, and Helen Wheatley. During the final stages of the project, I benefited from the encouragement of Toyin Falola and Robin Kilson. Robin's involvement in this project was especially meaningful to me: when I was a college student, she was the teacher who first got me interested in the history of colonialism. I am also grateful to Sean Culhane, the director of the University of Rochester Press, and to the press's two anonymous reviewers, one of whom subsequently chose to identify himself (Paul Richards) and one of whom did not.

My research on three continents would not have been possible without support from the Johns Hopkins University, in particular the Department of History, the Frederick Jackson Turner Society, and the Program in Atlantic History, Culture, and Society. A Fulbright Scholarship made it possible for me to spend 1992 in Mauritius, where the School of Social Studies and Humanities at the University of Mauritius hosted me. As I worked on revisions, I received support from the Department of Science and Technology Studies at Cornell University; the Training Grant in the Social Implications of Changing Knowledge in the Life Sciences of the National Science Foundation; and the Bernadotte Schmitt Fellowship of the American Historical Association.

I also wish to thank two archivists who were especially helpful: France Mayes at the Mauritius Chamber of Agriculture and Cheryl Piggott at the Royal Botanic Gardens, Kew. Other librarians and archivists at Cornell, Johns Hopkins, the Public Record Office, the Mauritius Archives, the Mauritus Sugar Industry Research Institute, the University of Mauritius, and the U.S. National Agricultural Library also helped me. Numerous farmers, sugar estate staff members, scientists, and extension officers shared information with me too, but they must remain anonymous.

I could not have completed this project without the help of my family and friends. Two people were especially kind: Joan Walker gave me a home away from home in England, while Vinesh Hookoomsing did the same for me in Mauritius. One person stands out among all the others in having encouraged this project. My wife, Joanna Miller Storey tolerated my extended absences from home and generally bolstered my morale. She also explored Mauritius with me, helped me to transcribe my notes, and edited several manuscript drafts. I dedicate this work to her in thanks for her support.

ABBREVIATIONS

CO—Colonial Office
COMDOA—Colony of Mauritius, Department of Agriculture
EPZ—Export Processing Zone
FSC—Farmers' Service Centre
ICS—Indian Civil Service
IMF—International Monetary Fund
MCOA—Mauritius Chamber of Agriculture
MMM—Mouvement Militant Mauricien
MSM—Mouvement Socialiste Mauricien
MSIRI—Mauritius Sugar Industry Research Institute
PMSD—Parti Mauricien Social Démocrate
POJ—East Java Experiment Station
PRO—Public Record Office, Kew
RBGK—Royal Botanic Gardens, Kew
RBGP—Royal Botanic Gardens, Pamplemousses
SRS—Sugarcane Research Station
VOC—Dutch East India Company

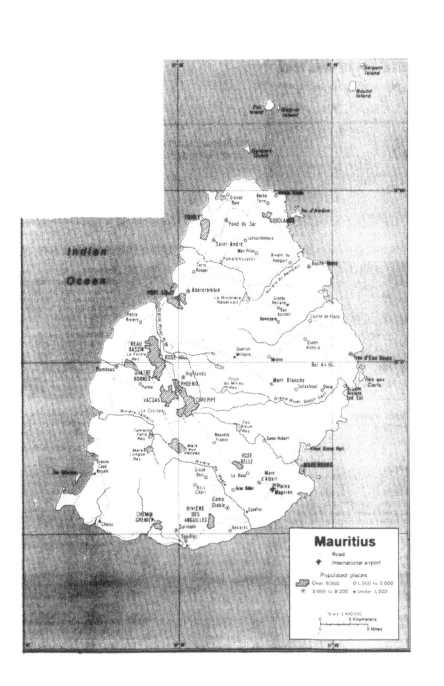

INTRODUCTION

SCIENCE, CANES, AND COLONIES

Historians of colonialism tend to neglect the history of science. This is a pity, because science played a fundamental role in the construction and perpetuation of many colonies. In this book, a case study of the Indian Ocean island of Mauritius, I will argue that science was important for colonialism. But this alone is not a surprising argument to make. It follows the pioneering studies of science, technology, and colonialism that appeared during the 1980s, especially the work of Michael Adas, Richard Grove, and Daniel Headrick. I chose to work on the history of science in Mauritius because I thought I might be able to amplify the findings of these scholars by examining the island's diverse cultural milieu, its long colonial history, and its rich archival sources.

Thanks to these sources I found evidence of complex local struggles over what should constitute the colony's science. These struggles involved peasants, planters, scientists, and officials; people from all ethnic groups; and people outside and inside Mauritius. Therefore, I will not only say that science was important in colonial Mauritius; that much is obvious. I will argue that Mauritians themselves influenced the course of colonial science. Mauritian history offers plenty of evidence that a colonial science, a colonial society, and a colonial state constructed and reinforced each other.

Historians of colonialism have missed a great deal by ignoring the history of science, but historians of science have also tended to ignore the history of colonialism. This book will bring the history of colonialism together with recent debates in the history of science. I start with the following premise: that science can best be understood in its social and cultural

1

context. Therefore, traditional historians of science will instinctively relegate this book to the category of "externalist" history. But I want to go beyond the shopworn debate between those who believe that science can best be understood by analyzing "internal" intellectual factors and those who give priority to "external" institutional, social, and economic influences. Historians of science must understand all aspects of science, its ideas, practices, and institutions as well as its place in society. As Steven Shapin argues, "There is as much 'society' inside the scientist's laboratory, and internal to the development of scientific knowledge, as there is 'outside.'"[1] Mauritius is especially interesting for social historians of science, not only because the island is a veritable Babel of languages and identities, but also because the island's economic history shows both rivalry and interdependency among sharply differentiated classes, all of whom produced scientific knowledge.

My argument about the mutual construction of science and colonialism stands in contrast to a handful of other works in this field. Most specialists would agree with me that science was important to colonialism, otherwise they would not write about it. But many of them subscribe to a theory of Eurocentric diffusionism, which says that Europeans imposed their sciences on the rest of the world. Debates about science and colonialism tend to pit conservatives against marxists, with controversy swirling around the question of whether the global spread of European science served the causes of enlightenment or oppression.

I will engage the ideas of other authors when it is appropriate, but mainly I think that the diffusionist framework has limited our understanding of science in the colonies. Diffusionism tends to reduce a rich historical record into simplistic binary oppositions: cores and peripheries, metropoles and colonies, oppressors and oppressed. This framework does have its advantages: if the central historiographical problem is to trace the influence of European science, then it is possible to rely exclusively on archival sources in Europe and the United States without bothering to consult sources in less comfortable places.

This book will employ sources from Mauritius, as well as from Britain and the United States, to relate a complex story about culture, politics, and science in a colony. The story takes place in sites both near to and far from the laboratories and lecture-halls of the European scientific and technological tradition. Science takes place wherever people produce knowledge about nature in a systematic and predictive way, and in Mauritius it was produced by many different sorts of people.

Recently, other writers have asked historians to pay more attention to the scientific discourses of Africans and Asians under colonialism. One of the best examples of this kind of work is James Fairhead and Melissa Leach's recent monograph on Kissidougou in Guinée. There, patches of forest can be seen on the savanna. From the 1890s until the 1990s, French scientists visited the area and concluded that the local inhabitants were converting forest to grassland, a view that was consonant with broader western views of African natural resource management practices: the patches of trees were remnants of old forests. But the French scientists never bothered to ask the villagers what was happening, and neither did anyone else. It was only in the 1990s, when Fairhead and Leach conducted some interviews, that the local people were given the chance to explain the patches of trees. Fairhead and Leach learned that all along, the villagers had been planting the trees in the grasslands. French scientists had been reading the history of the landscape backwards.[2]

But there is a key element missing from Fairhead and Leach's account, an element that is also missing from other work on the history of science, technology, and colonialism. If colonial natural resource practices were supported by evidence that local farmers found unpersuasive, then how did colonial scientists persuade officials and farmers to use them? What sort of rhetorical and social practices did they employ? Was it just a case of external authorities using brute force to make local subjects do their bidding? There are indeed many cases of colonial states forcing subject farmers to take up new agricultural practices. But generally speaking, colonial states did not have the resources to construct Orwellian apparatuses for social control. Instead, they relied on mixtures of local and metropolitan authorities and on blends of coercion and persuasion. This book on Mauritius defines science broadly as a way to open up the cultural, rhetorical, and social practices of scientists to scrutiny.

In keeping with such a broad notion of science, I have consulted a broad range of sources. Like most historians, I have used written materials, spending most of 1991 through 1995 reading articles, books, pamphlets, and reports. But I have also used what Simon Schama calls "the archive of the feet"[3] to get a sense of Mauritius, visiting with farmers, officials, and scientists over the course of nine months of research in 1992. I have tried to cast as wide a net as Paul Richards has done in his ethnographic and historical studies of Mende rice farmers in Sierra Leone. There, Richards found that local farmers have been conducting systematic selection trials on local wild rice varieties since before the first colonial agricultural officers appeared on the scene. He also found that "popular" science and "metropolitan" science co-existed and interacted in many ways at the center of the British Empire and also in such "marginal" places as Sierra Leone.[4]

Richards has focused his studies on farmers who grow rice, one of the most important colonial crops. My book focuses on the scientific research that supported another important colonial crop, sugar cane. I have drawn on historical scholarship on Europe's colonies, but I have taken much of my inspiration from recent work in the history of science. As Deborah Fitzgerald argues in her book about maize breeding in Illinois, "agricultural commodities" are ideal subjects for exploring the translation of theoretical knowledge into practical knowledge.[5] This history of agricultural science and sugar cane breeding in Mauritius follows Fitzgerald's lead, although I have treated a different crop in a different context. The history of sugar cane breeding in Mauritius shows that debates about politics, science, and society in a remote colony were quite important for setting both local and global research agendas.

To understand the relationship between politics, science, and society in Mauritius, one must first understand the sugar cane, the plant that has dominated the island's agriculture since the early nineteenth century. At various times, other tropical regions also supported intensive sugar cane monoculture, and many still do. Therefore, Mauritius shares much in common with the sugar islands of the Caribbean and the Pacific, as well as with the larger sugar-growing regions in Brazil, Louisiana, Natal, and Queensland.

By and large, historians and social scientists analyzing sugar-growing regions have taken the sugar cane plant for granted. To most of them, it serves as a leafy, green backdrop to a story of colonialism and coerced labor. In this view, sugar cane is merely a natural artifact, a biological organism whose existence is not to be questioned. People simply plant it, cultivate it, harvest it, and process it to produce sucrose. But as Bruno Latour would argue, this interpretation follows the modern supposition that a stark division exists between the worlds of nature and society. There is more to the sugar cane plant than meets the eye. It has a complex physiology and ecology. Substantial networks of producers and researchers have striven to prolong and improve its usefulness. The history of producing and planting sugar cane is filled with controversies that cross the boundaries of what is conventionally assumed to be the "political," the "social," and the "scientific."[6]

The sugar cane plant produces sucrose, and it has played a major role in tropical history, perhaps greater than any other plant. Over time, increasing demand for sweeteners spurred farmers to cultivate canes. In turn, the plant's botanical characteristics ensured the creation of particular methods of production. Scientists played an integral part in procuring and creating new kinds of cane, preserving the viability of both cane culture and the societies that relied on sugar production.

Over time people tinkered with canes to improve their yield. Farmers and scientists have selected and altered it substantially while holding political, social, and scientific objectives in mind. However, the sugar cane plant placed natural limitations on human intervention: scientists and farmers could only do so much with the material at hand. Scientists and cultivators interacted with the plant as part of a landscape that incorporated fields and laboratories. The sugar cane plant helped to shape the societies that tried to shape it. It did not determine history, but it was part of a seamless web in which polity, society, economy, science, and technology were interwoven.[7]

How did the sugar cane plant change over time, and how did the process of making new canes become stable? A complex network of scientists, farmers, politicians, officials, plants, tools, and pathogens interacted to produce and solve the problems associated with sucrose production. The interaction among the members of this complex network inspired fundamental debates, especially when different social groups sought to stabilize sugar cane breeding in different ways. Throughout the history of sucrose production, conflict has surrounded the production of knowledge about the sugar cane plant.[8]

To understand sugar cane science, it is necessary to understand the plant itself. Sugar cane is a large, perennial grass, of which there are five species. Three of these cannot survive without cultivation, making them completely dependent on humans, while the remaining two can grow wild. The most important species is *Saccharum officinarum*, which originated in New Guinea and is also called the "noble cane." During the eighteenth and nineteenth centuries, it was the most popular cane on European colonial plantations because of its large size, high sucrose content, and relative ease of harvesting. Two other cultivated species, *S. sinense* and *S. barberi*, were originally grown in Southeast Asia and northern India, respectively. They produce less sucrose than *S. officinarum*, but they grow more vigorously, suiting them to more extreme climatic conditions. The two wild species, *S. spontaneum* and *S. robustum*, have never been cultivated commercially because of their low sucrose content, but they happen to be vigorous growers. Of the four low-sucrose species, *S. spontaneum* has proven the most useful in hybridizing programs to improve upon *S. officinarum*. Since the 1930s, almost all canes in commercial cultivation around the world have been hybrid crosses of the noble *S. officinarum* and the wild species *S. spontaneum*.[9]

Hundreds of years ago, cane cultivation and selection began in New Guinea, where people still grow sugar canes for chewing and ornamentation. The New Guineans used careful observation and selection to discover and propagate the sweetest and most colorful varieties of *S. officinarum*.[10] Sugar cane reproduces sexually by making seed and tends to have a wide range of natural variation. It is also prone to mutations.[11] Cane growers can choose the canes that have the most desirable characteristics through patient observation of plants multiplying and mutating. Then, they can preserve newly found characteristics in these sexually reproduced varieties through asexual, vegetative propagation, also known as "cloning." It happens that cloning works particularly well in the case of cane, and simply consists of cutting a piece of cane, planting it in the ground, and waiting for roots and stalks to grow. Sugar cane cuttings have the advantage of being hardy so that they can remain viable over long sea voyages. Polynesian navigators took advantage of this trait and distributed the "noble" cane throughout the Pacific, bringing it as far as the Hawaiian Islands by the year 800.

The early expansion of European cane cultivation relied upon a very hardy and flexible cane. This was the first indication that cane selection could play an important role in the sugar industry. Today, this variety is known as the "Creole" cane, and it is thought to be a hybrid cross between two unknown varieties of *S. officinarum* and *S. barberi* that occurred in India.[12] (Technically speaking, a "hybrid" has parents from two different species, while a "variety" is a subdivision of a species. Nevertheless, I shall follow the convention of the world's sugar cane industries and refer to hybrids as varieties, except in so far as it is important to distinguish the two when discussing breeding.) Propagated by cuttings, the Creole cane first spread from India to China and Persia by the year 600, and from Persia throughout the Islamic world. During the Crusades, Europeans acquired a taste for sugar in the Levant, and they initiated the early Mediterranean sugar industry by planting Creole canes.

This variety served as the basis for the further expansion of sugar cane cultivation into the Atlantic. In 1420, Prince Henry the Navigator of Portugal sent it to Madeira, and from there growers brought it from the Canaries to São Thome off the west coast of Africa. On his second voyage, Columbus introduced the Creole cane to Hispaniola. During the sixteenth and seventeenth centuries, Europeans distributed it throughout the Americas. In 1737, the French introduced it to Mauritius, although in 1650 the Dutch had already introduced a noble cane variety to the island that was later known as "Otaheite."[13] But by the mid-eighteenth century, European cane growers in the tropics relied exclusively on the Creole variety, with the exception of the Dutch in Java.

Europeans created sugar colonies as part of one of history's most repugnant episodes in cross-cultural encounter. To produce a commodity used to sweeten foods and beverages, Europeans spread cane cultivation gradually from the Mediterranean, to the islands off the West African coast, and then to the Caribbean and Brazil. In doing so, they reduced indigenous populations by means of war, disease, and enslavement. Then they brought slaves from Africa to make up the resulting shortfall in labor, initiating one of the most horrific forced migrations in world history. Over time, European cane cultivation spread further throughout the Americas to islands in the Indian and Pacific Oceans and to Australia.

Large European sugar cane farms, called plantations, were among the first large-scale capitalist agricultural enterprises. They were cruel places where managers exercised quasi-legal jurisdiction over workers. As Philip Curtin argues, the most notable features of the sugar cane "plantation complex" were the widespread use of forced labor; the existence of populations that did not increase naturally; the integration of trade into long-distance networks; and the acknowledgement of the authority of remote European rulers. Wherever the plantation complex took hold, it transformed ecological systems and interrupted historical continuities in land, labor, and capital.[14]

Most historians, including Curtin, believe that during the nineteenth century, increases in metropolitan control and the emancipation of colonial slaves ended the plantation complex. By contrast, I wish to argue that the complex adapted to new conditions and survived. Capitalist sugar producers rationalized their operations and continued their involvement in local governance. Former slave owners turned to alternative methods of labor recruitment. At first, these were based on other forms of coercion, including indenture, tenancy, sharecropping, debt slavery, and "vagrancy" laws. Many of these methods had already been used during the earlier history of the plantation complex but were originally discarded because of the temporary advantages of slave labor. Small-scale cane farming replaced large-scale industrial plantations in many regions, but these small farmers still relied heavily upon larger sugar factories. Toward the end of colonial rule in many places, sugar producers adopted mechanized field techniques as well, eliminating much of their need for unskilled labor.

Despite changes in labor and a disappearance of large-scale production in many regions, a complex still exists. This is because, at its most fundamental level, the plantation complex was based upon knowledge of how to cultivate and improve the sugar cane plant, just as much as it was based upon a web of capital investment and exploitation of land and labor. Agricultural knowledge has always supported sugar production, but during the highly competitive nineteenth and twentieth centuries, it came to play an increasingly prominent role. As the old plantation complex was reeling from the effects of slave emancipation, the transregional and transnational exchange of plants and ideas came to flourish, and this was followed by the institution of experimental stations and research laboratories.

Knowledge of sugar cane botany has always given structure to the processes of sugar production. Sugar cane grows best in wet, warm tropical conditions, and yields the most sucrose when a warm, wet season is followed by a cooler, drier season. These climatic requirements restrict it to tropical and subtropical regions, which ideally do not experience frosts. Farmers cultivate sugar cane in many different parts of the world, but canes seem to produce the most sucrose at about 18° north or south latitude.[15]

Sugar cane botany makes special demands on field laborers. When the cane is ripe and ready for harvesting, workers must cut the cane close to the ground because the lowest part of the plant contains the most sucrose. They must cut carefully, leaving the stumps in good condition, because new canes or "ratoons" will grow from the old stumps and provide the next year's crop. Then workers must strip the sharp leaves and tops from the canes and hoist them into carts. Injuries from machetes and flying debris are not uncommon. Cane cutting is therefore difficult and unpleasant; it requires strength, skill, and agility.

Cane also has peculiar processing requirements that stem directly from its botanical properties. Soon after cutting, the cane's sucrose content diminishes, reducing the value of the crop. Producers must process canes into sugar as soon as possible, preferably within twenty-four hours. Processing "on the spot" also reduces transport costs because sugar canes are heavy and cumbersome, while the extracted sugar has more value for the same unit of bulk. Early sugar plantations had their own factories to ensure rapid and reliable processing. Toward the end of the nineteenth century, railroads and new factory technologies in many regions made it more efficient to draw canes from greater distances to larger central factories. Depending on the region, factories often kept their own cane lands in production, even as they purchased canes from small-scale producers. But any savings the central factories made in reducing fixed costs and labor costs had to offset the increase in transaction costs that subcontracting for a cane supply entailed. These depended on local laws and economic and social conditions. In many parts of the world, the vertically integrated, large-scale sugar cane factory-farm remains the preferred method of production.

Sugar factories dictate a sporadically intensive labor regime. Factories must process canes when they contain the most sucrose. During the harvest period, which can last several months, factories typically operate around the clock. To ensure the factory always has enough cane, field laborers must work long hours to harvest as much as possible. After the harvest in most cane-producing regions, laborers replant the fields of the oldest ratoons, which is also heavy, skilled labor. But afterwards, there are only light tasks to perform such as weeding, and it is not economical for a large-scale factory-farm to employ the full contingent of workers used during the harvest.

This production schedule poses a problem for sugar producers: how to ensure a ready supply of labor during each year's harvest, when it is to the laborers' advantage to find steady, year-round employment, rather than six months of back-breaking work followed by six months of comparative idleness. Late twentieth-century sugar producers have solved this problem through mechanization, complex labor contracting arrangements, migrant workers, and diversification into other crops and businesses that provide work during the "inter-crop" period. Today, some governments even require large cane farms to provide year-round employment. But during the earlier centuries of sugar cane production, plantations throughout the world resorted to coercion to keep labor available.[16]

Sugar cane botany explains, in part, the plant's historical connection with colonialism and coerced labor. This does not excuse colonial exploitation, nor does it mean that plants determine human history. But those who improve the cultivation of sugar cane (or any crop, for that matter) are in a position to make the most money. Scientists who study the improvement of sugar cane varieties and inform the public about cultivation practices play a significant role in production. Economies came to depend upon sugar cane for geographical as well as political reasons. The nature of the sugar cane plant itself, together with changes in the systems of world sugar production, led cane growers to depend upon scientists. Nevertheless, debates about the best ways to grow cane took place in the shadows of history. One Mauritian cane scientist remarked to me that, "we work in the dark." It is surprising, in fact, that these people have gone relatively unnoticed. The producers who were able to master the sugar cane gained considerable advantages, while those who did not found other forms of support, or failed.

This book illustrates changes in human understanding of the sugar cane plant, showing this global process in Mauritius. It is an investigation of the efforts of agricultural scientists as well as farmers and state officials to sustain and develop an agricultural colony. The evolution of a cadre of scientists capable of improving the sugar cane plant caused and reflected fundamental changes in the relations between state and society. I hope that this study of agricultural science in Mauritius will shed light on the workings of colonial capitalism and agrarian societies, as well as on the history of science and colonialism.

One must grasp the connections between global and local political struggles in order to understand the production of imperialist science and technology. It would be wrong to diminish the historical importance of researchers in Europe. It is still important to understand how metropolitan institutions influenced the training and outlook of colonial scientists. However, historians must place their work in a wider context that takes struggles in the colonies seriously.

This book goes beyond Eurocentric diffusionism to examine the ways in which various people struggled to understand nature in the colonial context. It examines the history of science and technology in Mauritius, Britain's premiere sugar-producing colony, from a comparative perspective. Sugar cane breeding and other aspects of crop research reflected patterns found throughout the British Empire and other parts of the world. Therefore, the book weaves together the history of agricultural science with the political, social, and economic history of Mauritius, stressing the interactions between internal and external historical forces. Without scientific research in plant technology, the kind of colonialism associated with sugar cane cultivation would not have been possible. Scientists were not impartial bystanders, but participated in colonial political, social, and economic discourses. They produced the knowledge and tools that perpetuated the colonial economy. They also implemented decisions on how to manage natural resources. In Mauritius, colonial politicians and agricultural scientists were partners in a symbiotic relationship.

1

CONSTRAINTS, COLONIZATION, AND CANES, 1500–1853

Mauritius is a small, pear-shaped island in the southwest corner of the Indian Ocean. It is one of the few countries in the world where the entire population can trace its origins to some form of colonial migration, including settlement, enslavement, and servitude. There are not any "native Mauritians," but there are, instead, a million Mauritians who are descended from African, Asian, and European migrants. Mauritius owes its existence as a nation so completely to the former colonial powers that its history ought to show the starkest forms of colonial exploitation. Frequently, it does. But the history of Mauritius is not just a story of migrant laborers and their descendants losing unequal battles with European colonizers.

Geography and the Early History of Mauritius

Mauritian history is about people struggling to overcome the constraints of geography. The first time I stood on top of one of the island's small mountains, the Trou aux Cerfs in Curepipe, I was struck by the island's small size. I could turn in a circle and survey the entire island, which is slightly smaller than the state of Rhode Island. All around, dark, jagged mountains rise up from the flat, green plains, a sure sign that the island was formed relatively recently by volcanoes.

Insularity provides an important constraint, but so does isolation. Looking beyond the black mountains, the green plains, and the white beaches, I was struck by the vastness of the southwest Indian Ocean. The closest lands are not even visible, and these, the neighboring Mascarene Islands of Réunion and Rodrigues, are also quite small. Large land masses are even farther away. It is 800 kilometers to Madagascar, 2,000 to Mozambique, 4,000 to India, and 5,000 to Australia.

Isolation and insularity are obvious features of the Mauritian landscape, but topography and climate impose further geographical constraints. Mauritius lies within the tropics, but the southeast trade winds blow throughout the year, ensuring that the thermometer stays on the cooler side of the tropical range. On any given day, temperatures can be sweltering in coastal Port Louis and cool in highland Curepipe. The winds bring considerable rainfall to Mauritius, especially between December and April, but storms usually last only an hour or two. In the drier, leeward part of the island it seems sunny most of the time, but on the windward slopes of the central plateau it often seems cool and rainy. In 1896, Mark Twain spent ten days on the central plateau in soggy Curepipe, which led him to complain that Mauritius "is the only place in the world where no breed of matches can stand the damp. Only one match in sixteen will light."[1]

Mauritian rainfall can dampen the spirits, but it definitely imposes important constraints on agriculture. The island's soil is generally fertile, but land in areas with high rainfall and humidity has lost soil nutrients and tends to be too acidic for most crops.[2] Even so, most Mauritian rainstorms are mild affairs. But during the summer months an average of four hurricanes called "cyclones" form in the Indian Ocean. When they pass at a distance from Mauritius they bring beneficial rainfall, but when they pass directly over the island, as they do once every two or three years, they cause immense destruction. The most powerful cyclones destroy crops and trees as well as flimsy structures. They discourage settlement and agriculture.[3]

At first Mauritius did not have any human inhabitants, and much confusion has surrounded the island's discovery. Cartographic evidence suggests that Arab navigators knew of the Mascarenes and that they called Mauritius Dina Katab, the "abandoned" or "devastated" island, perhaps because they found it ruined by a cyclone. During the 1510s, the Portuguese were the first Europeans to explore the southwest Indian Ocean, and their cartographers named Réunion, Mauritius, and Rodrigues the Mascarene Islands after one of their captains, Pedro Mascarenhas. The Portuguese do not seem to have used the islands for much of anything, except possibly as emergency stopovers on their route to the Indies.[4]

For the first time, the Mascarenes were no longer completely isolated, and they even lay near important European trade routes. Because of this, English and Dutch traders began to show a moderate interest in the Mascarenes. In 1598, the Dutch East India Company (VOC) claimed Mauritius, naming it after the Dutch head of state, Maurice of Nassau. The company thought that the island had better harbors than the other Mascarenes, giving it some potential as a refreshment station for their ships on the return voyage to Amsterdam from the East Indies. Even so, between 1601 and 1638 only a handful of VOC ships stopped there, and Portuguese, Danish, French, and English ships also used the island.[5] Europeans began to fell the island's ebony trees for timber, the first attempts to exploit Mauritian natural resources. In 1638, the VOC decided that the island's ebony forests were a resource worth monopolizing, and sent twenty-five settlers to cut wood, explore, and scare off foreigners. This was the first in a string of Dutch efforts to create a permanent population on the island, all of which failed miserably. The island was fairly remote, and its internal geographic conditions were largely unknown.

The earliest settlers probably had difficulty with the peculiar weather and remoteness from markets that have always constrained Mauritian agriculture. In 1639, they introduced the sugar cane plant to the island and used its juice to distill rotgut liquor. Canes resist cyclones better than other cash crops that the Dutch tried, such as tobacco and indigo, but the development of commercial agriculture was not really possible without a larger and more stable supply of labor.[6] There were no "native Mauritians" to exploit, and all laborers had to be imported, either through enticements to settle or through forced removal. To the first Dutch settlers, it may have seemed that Mauritius was better-suited to trade rather than farming.

The first Dutch colony in Mauritius lasted only twenty years and never amounted to very much. In 1658, the VOC abandoned the struggling colony, but in 1664 it returned again to begin another series of failed settlements. The company had probably given up any hopes for a prosperous island economy, but it still wanted to keep the island out of foreign hands because of its trees and harbors.[7] By the early eighteenth century the colony had once again begun to fail. In 1710, the Dutch evacuated the island for the last time.

Some historians have blamed the Dutch failure on inept administration, but it seems that social problems and geographical constraints also contributed to the colony's demise.[8] Given the problems the Dutch experienced in persuading settlers and slaves to remain on farms, labor was not likely to become available without subsidies from the VOC. To make matters worse, at least six major cyclones struck the island during the Dutch occupation, destroying houses and crops.

The Dutch may have had little long-term effect on the peopling of Mauritius, but they definitely had a negative influence on the island's environment. The colonists stripped the forests of ebony until the 1650s, when the European market became saturated. They also killed off the large, flightless dodos, who had never before encountered a natural predator. When the Dutch whistled for them, the unsuspecting birds waddled over, only to be slugged over their heads. By 1681, there were no more dodos on the island, but today the bird lives on as the extinct, flightless national mascot. By 1700, the Dutch also finished off other unusual indigenous species such as the giant land tortoises, the sea turtles, and the dugongs. To make matters worse, the Dutch introduced deer, rats, and monkeys to Mauritius, all of which have plagued island farmers for more than three hundred years.[9]

At first glance, then, the Dutch settlement of Mauritius may seem to be an irrelevant episode, especially if one blames the colony's failure on the ineptitude of the colonists and their administrators. Nonetheless, the episode is instructive because it illustrates how geography is one of the fundamental problems of Mauritian history. Climate constrained cultivation, while the island was remote from administrative centers and markets.

Creating a New "Ile de France," 1715–1803

The departure of the Dutch from the southwest Indian Ocean left a vacuum filled most easily by the French, who were actively colonizing the region. In 1642, while the VOC was cutting ebony in Mauritius, the French Eastern Company was taking possession of nearby Réunion, naming the island "Bourbon." The French East India Company assumed the Eastern Company's rights, and in 1665 it sent its first party of colonists to Bourbon. The island colony made slow progress because of the company's continued focus on colonizing Madagascar. Bourbon was just as isolated and insular as Mauritius, but to make matters worse, it lacked a harbor and had no readily exploitable natural resources.[10]

Nevertheless, the Mascarene Islands were increasingly becoming subject to the whims of European markets. As in other colonies, seemingly insignificant changes in European tastes and conditions could introduce profound changes in the management of local natural resources. This is what happened in the early eighteenth century, when the French East India Company took an interest in coffee cultivation and thereby changed the fortunes of both Bourbon and Mauritius. In 1715, a company expedition to the Red Sea port of Mocha obtained coffee plants for Bourbon. On the return trip, the French traders' ships took a detour and laid claim to Mauritius.

Coffee grew well in Bourbon and became the island's principal crop for the rest of the eighteenth century, in spite of competition from the French Caribbean.[11] But for some time, the company's intentions toward Mauritius remained unclear. In 1720, the company sent a small party of Bourbon's settlers to the island, which they renamed "Ile de France." During the next year, it sent a larger group of settlers directly from France. They occupied Vieux Grand Port on the southeast coast, the site of an old Dutch settlement, but their efforts to colonize Mauritius were scarcely more successful than those of their Dutch predecessors. The settlers' first attempts at agriculture met with frustration, mainly because droughts, rats, deer, and monkeys destroyed the crops. In 1723, the governor reported harvesting a small amount of tobacco, maize, and rice, but the rats made it difficult to store any food. In fact, the rat problem was so bad that a visiting French missionary re-christened the island the "kingdom of the rats." Every morning, he claimed, the colonists woke up and compared the rat bites they had received during the night. The settlers were forced to rely on deer-hunting and the company's ships for their food.[12]

Once again, climate and isolation were taking their toll. Between 1725 and 1735, the situation in Ile de France deteriorated, despite numerous changes of administration and the arrival of about one thousand settlers. Cyclones ravaged the settlement periodically. Agriculture flopped completely, except for a few anemic coffee plots. Then in 1729 the company granted Bourbon a monopoly on coffee production, and what remained of the Ile de France coffee plots had to be ripped out. That year, the Provincial Council of Ile de France noted that the colonists would most certainly starve if they did not receive outside assistance.

The colonists were a rum lot, too, recruited from the dregs of Breton and Bourbon society and "protected" by a garrison of unsavory Swiss mercenaries. Not surprisingly, the colonists earned a reputation for sloth, anarchy, and drunkenness, made worse by their remoteness from an administrative center. They lacked enterprise and, according to a visitor, "awaited ships from India like Jews waiting for the Messiah."[13] One of the only positive accomplishments of the island's government was to shift the principal settlement from Vieux Grand Port in the southeast to Port Louis in the northwest. While the prevailing southeasterly trade winds pinned ships into Vieux Grand Port, they made it easy to sail out of Port Louis. Escape was preferable to arrival.

During the 1720s, the company imported slaves to redress the persistent labor shortage. The French were experiencing one of the same constraints that the Dutch had encountered, namely, the absence of any "native Mauritians" who could be forced to do the dirty work. The company turned to slave merchants to provide the labor. The merchants obliged, shipping slaves from West Africa, Madagascar, East Africa, and India at the rate of about six hundred per year.[14] While working for their French masters, the slaves developed a pidgin rooted in French and a mixture of African languages. This eventually became Mauritian Kreol, which islanders speak today; the language is related to the Creoles of Réunion, the Seychelles, Haiti, Martinique, and Guadeloupe.[15] In 1731, given the high transportation costs and mortality rates involved in shipping West Africans to the Indian Ocean, the company began to show a preference for slaves from Madagascar and East Africa, so that these slaves began to hold a predominant position in the slave culture of Ile de France.[16] In any case, many slaves ran away into the forest and formed communities of maroons.

During the late 1730s, the fortunes of Ile de France began to improve. Traditional accounts attribute major changes to the arrival of Bertrand François Mahé de Labourdonnais, a wealthy Breton squire and sea-captain who served as governor of both Ile de France and Bourbon from 1735 until 1746. Labourdonnais supervised the construction of fortifications and a shipyard, and he also formed a naval service for the Mascarene Islands.[17] But there was a down side to his activities, too. Labourdonnais's construction projects in Port Louis, together with his encouragement of agriculture in both islands, caused the demand for slaves to soar. The annual rate of slave imports into the Mascarenes rose to 1,200 or 1,300 people, most of whom came from Mozambique. After 1745 the trade dwindled, only to pick up again in 1749, when the rate rose to 1,300 or 1,400 per year.[18]

Even with Labourdonnais's efforts to build a port and to import slave laborers, the befuddled policies of the French East India Company hindered most efforts at development. The company kept its monopoly on the islands' trade, except for an experimental free-trading period between 1742 and 1747. The company abused its authority, selling supplies to settlers at inflated rates while paying rock-bottom prices for colonial produce. The remote islands were also chronically starved for currency, encouraging speculation in land and goods. If that was not bad enough, persistent droughts and cyclones discouraged colonists from farming.

Up until the mid-1760s, Labourdonnais and his successors struggled to establish gentleman-farmers in the Mascarenes. Labourdonnais himself recognized the importance of natural resource management to the colony's future. He experimented with crops and encouraged coffee production in Bourbon, as well as food production in Ile de France. He was personally responsible for introducing manioc from Brazil. He hoped that this crop, along with maize, would meet the nutritional needs of the burgeoning slave population, who would otherwise be quite isolated from substantial food supplies.[19]

Labourdonnais was also shrewd enough to recognize the potential for sugar cane cultivation on Ile de France. The plant tends to resist cyclones better than other crops, and it grows well in the island's tropical climate. Labourdonnais ordered the island's first primitive sugar factory equipment, and by the late 1740s, two factories were operating. Like sugar factories in the New World, they consisted of three vertical rollers driven by water or animal power, which drained the sugar juice into a series of copper boiling pans called a "battery."[20] One of these can still be seen at Pamplemousses, the site of Labourdonnais's estate.

But the efforts of Labourdonnais and later governors to promote agriculture failed against the persistent geographical constraints. Isolation and weather, blended together with European domination, all ensured that Ile de France became more of a trading post than an agricultural colony. European trade with India was growing. During France's wars with Britain, Port Louis became an important naval base. The settlers preferred to invest in privateering and shipping rather than farming, partly because of the rampant land speculation and currency fluctuations that plagued the island's economy. Other troubles also encouraged trade. When France lost most of its Indian colonies in 1763, the French East India Company spiraled into bankruptcy. The French government assumed the direct administration of the Mascarenes, imposing a military governor-general and an economic-policy intendant upon the local inhabitants. The Crown permitted free trade, which transformed Port Louis into a boom town.[21]

The transition to Crown rule also had significant implications for natural resource management in Ile de France. This was largely due to the influence of Pierre Poivre, who served as the intendant between 1767 and 1772. Poivre was one of France's most famous natural philosophers, and it was under his leadership that statecraft became even more closely intertwined with agricultural science and environmental policymaking. Richard Grove argues that Poivre treated Ile de France as an experimental proving ground for radical physiocratic principles. He followed François Quesnay, who stressed the importance of agriculture for the economy, and who also believed in the connections between agricultural, medical, and moral health. Poivre was also influenced by Richard Cantillon, one of the first economic thinkers to criticize mercantilism from the standpoint of universal economic principles. Cantillon supported the interests of small farmers over large landowners and he also believed that the state should support agriculture by providing scientific research and public works.[22]

These ideas resonated with Poivre's own botanical research and also with what he observed during his prior adventures in Africa, Asia, and the East Indies. When he became intendant in Ile de France, he put physiocracy into action. He did so through persuasion and executive orders, ariculating his program in speeches, letters, and edicts. He tried to build what Grove calls a "moral economy of nature," a political program that included forest conservation, land reform, and botanical research. Poivre spoke out against the land speculators who were despoiling the forests. He encouraged the production of new food crops and cash crops. He also opposed slavery, which he considered to be an abomination against the natural, moral, and economic order. But by the time he left Ile de France in 1772, Poivre and his physiocratic supporters in Paris and Ile de France were finding themselves outmaneuvered by mercantilist governors and ministers who were more interested in short-term profits than long-term prosperity.[23]

Poivre's utopian vision did not include slavery, but free trade stimulated slave imports. As Poivre realized, the growth of slavery was having profound social and moral consequences. A three-tiered society was beginning to emerge, composed of a thin veneer of Europeans and mixed-race freedmen benefiting from the work of a slave population that was predominantly African in origin. Unfortunately, little is known about the social history of the slaves of Ile de France. The French imported a total of approximately 160,000 slaves to the Mascarenes, 110,000 of which arrived between 1767 and 1810. By the end of the *Ancien Régime*, 1,000 slaves were reaching the islands each year from Madagascar alone, while between November 1786 and January 1788, traders landed 3,000 East Africans in Port Louis alone. They were a diverse group: approximately 45 percent came from Madagascar, 40 percent from Mozambique and the Arab trading posts along the East African coast, 13 percent from India, and 2 percent from West Africa. The Africans tended to work in the port, in construction, and in agriculture, while the Indians tended to work as artisans and domestics. A numerical preponderance of males over females, along with poor working conditions, ensured that the slave population of Ile de France did not grow naturally.[24]

Local laws designed to control the slaves provide some insight into their social condition. The Code Noir of 1723, supposedly a humanitarian measure, spelled out in great detail the legal restrictions upon slaves in Ile de France. Among other things, it prohibited interracial marriage or co-habitation and restricted the property rights of freed slaves. It also codified the punishments that masters could mete out to slaves; runaways, for example, could be chained, whipped, branded, mutilated, hamstrung, or killed, depending on whether it was their first, second, or third offence. Today a myth persists that slavery on Ile de France was "kinder and gentler" than most places, but the Code Noir suggests colonial bloody-mindedness. After 1773, about 10 percent of the slave population absconded during any given year. Considering the frightful punishments for running away, this figure suggests that conditions were less than ideal.[25]

Slaves provided most of the labor on Ile de France, but free non-Europeans played an important role, too. During the early years, runaway slaves joined maroons in the forest. Later, they sheltered among the emerging "free colored" population of Port Louis. Mostly these *gens de couleur* were free Indian artisans and laborers, as well as manumitted slaves, concubines, and their descendants. In 1800 they made up 7 percent of Ile de France's population, and by 1820 they were beginning to achieve some social and economic prominence.[26] Many of them participated in the maritime economy and in small-scale food production.[27]

While the slaves and freedmen worked, most of the island's European population led a comfortable life. Most were French merchants, sailors, and soldiers. They settled permanently on the island, by contrast to the Europeans in the Caribbean who tended to become absentee proprietors. As a result, many of today's Franco-Mauritians can trace the arrival of their ancestors back 250 years or more. On the whole, European life in late eighteenth-century Ile de France probably compared favorably with life in France itself, although some *petits blancs* were not much better off than the *gens de couleur*.

Free trade and isolation ensured that even the political disruptions of the French Revolution did little to change Ile de France's basic orientation to the sea. At first, the island's merchants and settlers welcomed the opportunity to sack the royal administration, but in 1794, when they learned that the new republic had abolished slavery, they sacked the revolutionary government, too, and severed all political ties with the mother country. The islanders then took advantage of the state of war between England and France. They took up privateering, attacking English shipping in the Indian Ocean while still flying the French tricolor as a matter of convenience. The privateers' loot attracted even more neutral trade to Port Louis than usual, increasing the town's surface prosperity. The French Revolution also had little effect upon the slave trade.[28]

Nevertheless, prosperity was illusory. The economy remained unstable because reserves of the main currency, the Spanish dollar, were dwindling along with supplies of the local paper currency. This situation encouraged speculation in land and goods, undermining agriculture at a time when war threatened to cut off the island's external food supplies.[29] In 1803, the islands' governments agreed to reintegrate themselves with metropolitan France after Napoleon made it clear that he would allow slavery to continue. Local civil and military administration reverted to pre-1790 norms, but the new government also imposed the Napoleonic Code, stopped printing paper money, and established a school system. But even with these accomplishments, the Napoleonic era was to be brief.

Sugar Cane and the Scientific Tradition of Ile de France

During the late eighteenth century, the Ile de France's farmers finally began to find a reliable market for a crop: sugar cane. French merchants purchased land on Ile de France as insurance against unstable currency. During the wars with Britain, they hedged their bets on trade by trying to make their land turn a profit. Sugar cane grew best, and during the Napoleonic Wars neutral ships bought Ile de France sugar and sold it in Europe, which was then deprived of the produce of Saint Domingue. Conveniently, sailors calling at Port Louis also had an unquenchable thirst for cane liquor.

When the Ile de France's farmers expanded cane production during the late eighteenth century, they had no idea that the sugar cane plant would cover the landscape for the next two centuries. The plant did not determine the island's history, but the choice to cultivate it had significant ecological, economic, political, and social consequences. To understand these fully, it is necessary to understand how farmers selected sugar cane varieties.

The Creole cane yields good quality juice on a reliable basis, but it also contains a high proportion of fiber. This problem inspired producers to acquire imported cane varieties, particularly the so-called "noble" varieties from the Pacific, which had more sucrose.[30] The French explorer Bougainville probably made the first deliberate introduction of a higher-yielding variety to replace another cane. During his voyage of 1766–1768, he acquired the "Otaheite" cane in Tahiti. On his way home he introduced it to Ile de France and Bourbon, but he was not aware that more than a hundred years earlier, the Dutch had already introduced this same variety, although it was not widely cultivated.[31]

The substitution of Otaheite for Creole canes proved successful, and more people became interested in the potential of new varieties to increase production. But this development must be placed in a broader intellectual context. Madeleine Ly-Tio-Fane has written extensively about how, during this period, many residents of Ile de France were taking a strong interest in botany and natural history. These included some of the large landowners who dominated the nascent sugar industry, but they were not the only ones. Despite the island's isolation, Ile de France intellectuals developed ways to teach and discuss botany and natural history, both locally and through connections with international networks. During the last decades of French rule, world-renowned naturalists visited Ile de France, including Commerson, Baudin, Bory de Saint Vincent, Michaux, and Aubert du Petit Thouars, while several botanically minded colonists made international reputations for themselves. There was even enough local interest in natural history to form two learned societies.[32]

The vibrant intellectual life in Ile de France is interesting when it is viewed in light of recent scholarship in the history of science and colonialism. The pioneering work in the field was done by George Basalla, who believed that the history of colonial science and technology was mainly a story of the diffusion of western science out into the colonial periphery. Basalla elaborated a three-stage model for scientific progress in the colonies that did not consider the possibility that Westerners might acquire scientific knowledge from non-Westerners.[33]

Basalla's diffusionist model continues to influence scholarship on colonial science and technology. One of the most prominent historians in the field, Lewis Pyenson, has written extensively about the diffusion of European astronomy, geophysics, and meteorology to Dutch, French, and German colonies. According to him, even in colonies far away from Europe, there was "no colonial stamp" on the work of astronomers, geophysicists, and meteorologists.[34] Therefore, Pyenson believes that the history of colonial science demonstrates the internalist thesis: science can flourish without cultural and social influences.

This distinction between pure and applied sciences is both arbitrary and convenient.[35] Pyenson deliberately ignores agronomy and engineering because he believes that these "practical discourses" are too tainted by money to be considered pure sciences. Pyenson prefers to write about the loftier achievements of physical scientists, who seem to him to be above social influences. But agronomists and engineers are scientists. They employ systematic methods to understand and influence the natural world. It is also impossible to separate the "internal" methodologies of these scientists from their "external" social and cultural practices. It may be the case that Western scientific methods were exported wholesale to the colonies, but their inherent worth was not necessarily enough to persuade colonial people to take them up. Metropolitan methods had to predict natural phenomena in a systematic way, but the importers of these methods also had to use social and cultural means to achieve credibility.

The diffusionist thesis is overly simple, but it does have some support. France's most important sugar colony, Saint Domingue, had a vibrant intellectual life, and a comparison with Ile de France is instructive. In Saint Domingue, science was dominated by metropolitan institutions and did not have strong local roots. James McClellan argues that the French state sponsored scientific and medical research in Saint Domingue as part of a broad, sustained effort to develop the colony. Government scientific institutions played a large part in this task, as did Parisian and provincial scientific societies. But when the French pulled out of Saint Domingue, Haiti found itself "beyond the pale of Western science." McClellan examines many sources and concludes that "science and society in colonial Saint Domingue cannot be considered apart from contemporary science and scientific insitutions in France . . . colonial science in Saint Domingue, while on the periphery, was anything but peripheral."[36]

McClellan's findings on Saint Domingue raise important questions about science and colonialism, especially when they are compared with Ly-Tio-Fane's research on Ile de France. McClellan shows that in Saint Domingue, the metropole dominated science. His judicious consideration of a broad range of sources supports the diffusionist model of colonial science more convincingly than the dogmatic efforts of Pyenson. By contrast, Ly-Tio-Fane shows that a local scientific tradition took root in Ile de France, even as local scientists continued to engage with ideas from the wider world. Local science flourished despite wars, revolutions, and changes in administration, a situation that contradicts the diffusionist model.

Richard Grove's research in both metropolitan and colonial archives tends to amplify Ly-Tio-Fane's findings on the importance of science in the colonies, and he, too, contradicts the diffusionist model. Grove sees the origins of European environmentalism in the complex interaction between metropolitan scientists, colonial officials, colonial settlers, indigenous non-European peoples, and colonial ecosystems. He emphasizes the importance of newly acquired island colonies like Ile de France in the formation of a global discourse about deforestation and dessication, focusing on the relationship between indigenous notions of natural resource management in the colonies and the emergence of a new global discourse about the environment.[37]

The literature on colonial science is not yet sufficiently developed that one can say, with any certainty, whether the diffusionist model or the local model was more typical. But at the very least, the cases of Saint Domingue and Ile de France do support the argument that colonial political and social development were closely bound up with debates about scientific research, debates that took place both in the metropole and in the colonies.

Thanks to the efforts of Ly-Tio-Fane and another Mauritian historian, Alfred North-Coombes, we know that even as Ile de France passed in and out of the control of metropolitan France, a number of islanders maintained a strong interest in sugar cane botany. Not every sugar planter was an enthusiast; in fact, the lack of finance capital made some of them decidedly conservative. But the fortunes of the island were becoming closely bound up with the sugar cane plant, which stimulated greater interest in research and collecting. One of the most prominent local naturalists, Cossigny de Palma, also cultivated sugar cane. He was one of the first to recognize the importance of importing new varieties. In 1782, he introduced the Bamboo and Gingham canes from Java. After seven years of trials, he distributed them to the Pamplemousses gardens and several fellow planters. Cane trials still flourished, but the Creole and Otaheite canes remained the principal varieties.[38]

The French government recognized the value of Cossigny's experiments and exported the Otaheite cane and some of his other canes to Martinique. By 1793, the Otaheite was exceeding all other varieties in the New World. It spread throughout the French Caribbean under the name "Bourbon," because growers thought it came from Bourbon and not Ile de France.[39] During his second voyage in the Pacific, Captain Bligh also acquired the Otaheite (Bourbon) cane in Tahiti. In 1793, he introduced it to Jamaica, along with three other varieties that did not take hold. By the first decades of the nineteenth century, Otaheite cane grew extensively in most of Europe's sugar colonies, despite a Dutch attempt to introduce several other noble cane varieties from Java to their Caribbean colonies.[40]

These late eighteenth-century efforts to replace the Creole cane with noble canes initiated a pattern of cane experiments that lasted until the end of the nineteenth century. First, New Guineans selected and cultivated noble canes for their decorative and gastronomical qualities. Through the process of selection, they transformed wild canes into domesticated plants. Then European visitors to the Pacific acquired the canes, transported them to Mauritius, and planted them in experimental plots to test and demonstrate their productive capabilities. Like the New Guineans, Mauritians also believed that they were acquiring plants from the wild. In fact, the New Guineans had narrowed the selection for them. This pattern of cane acquisitions reflected the state of European natural history at the time. Investigators scoured the fields and forests of the wider world, collecting useful species and removing them to European gardens. The cane's botanical characteristics, especially its portability, helped people to disseminate it from the gardens of New Guinea to the plantations of the Southwest Indian Ocean and the Caribbean. The plant's properties influenced the changes in the landscape that were associated with colonialism.

Even in the late eighteenth century, sugar cane growers depended upon the intelligent selection and propagation of good varieties. Mauritius had the right kind of tropical climate to support sugar cane cultivation, but it also had the right kind of intellectual climate. Over time, even as the island's economy and polity changed, the Franco-Mauritian planters continued to influence agricultural research, usually by scrutinizing research institutions and sometimes by conducting their own research. This contradicts the diffusionist model of colonial science, in which the local cultural context for agricultural research is insignificant. It tends to confirm Paul Richards's argument that local planter influence ensures the most successful agricultural research systems.[41]

When viewed from a comparative perspective, the emergence of a local scientific tradition may have helped Mauritius to avoid the kinds of problems that other former sugar colonies have experienced. According to James Bartholomew, non-Western societies with existing scientific traditions have adapted most rapidly to the twentieth-century West's predominant scientific methods.[42] Over the long term, local scientific traditions have helped Franco-Mauritian sugar planters, and although they were not "non-Western" they were colonial subjects in a remote location. However, Mauritius became so socially diverse during the nineteenth century that the history of science and development there makes it difficult to employ simple categories like "West" and "non-West" and "colonizer" and "colonized."

The Advent of British Rule

Since the eighteenth century, Mauritius has been subject to the whims of external markets and rivalries. These have brought major social changes to the island, and they have also involved a diverse range of classes and cultures in sugar cane cultivation. The nineteenth century witnessed some of the most dramatic changes.

The Napoleonic Wars reaffirmed the strategic usefulness of Ile de France. Privateering continued to be highly profitable, but in 1806 it provoked the British navy into blockading Port Louis. At first trade suffered greatly, and before long it seemed that French rule in the Mascarenes was about to come to an end.[43] In July 1810, a British squadron captured Réunion without firing a shot. (During the Revolution, Bourbon's colonists had changed the island's name to Réunion.) The next month, the French navy trapped and destroyed a British squadron off the southeast coast of Ile de France. This feat was one of Napoleon's only major naval victories, and it is duly inscribed on the Arc de Triomphe, but it did little to stem Albion's tide. The Royal Navy returned in December 1810, met with little resistance, and captured Ile de France.[44]

The 1814 Treaty of Paris gave Great Britain permanent control over Ile de France, which was promptly renamed Mauritius. The British also gained control of the sparsely populated island of Rodrigues. The British had hopes of using Port Louis as a trading post, just like the French had done. However, they had no interest in harborless Réunion, which they returned to the previous owners.[45]

At first the British introduced only minor administrative changes to Mauritius. According to the terms of the French capitulation, the British agreed to respect the laws, customs, religion, and property of the local inhabitants. They made Mauritius a Crown Colony, placing it under the rule of a governor who was supposed to make most of the important decisions himself. After 1825 the Colonial Office instructed him to consult with an appointed Council of Government, composed of the colonial secretary, the chief justice, the controller of customs, and the officer commanding the troops.

Despite their broad administrative powers, the early governors of British Mauritius were weaklings who afforded the Franco-Mauritian elite plenty of room for political maneuver. The French administrative and judicial system was still intact, thanks to the terms of the capitulation. The British had little experience administering colonies where the Napoleonic civil and commercial codes were in force (although the Napoleonic penal code had not yet been adopted at the time of the surrender). A maze of pre-Napoleonic laws still existed too, and the colonists interpreted these to their own advantage. The court system was also large and unwieldy, considering the small size of the island. A clique of Franco-Mauritian judges and lawyers retained considerable power to obstruct British administrators and creditors, particularly under weak governors who could not penetrate the insular culture.[46]

In many ways the British government preferred to keep a certain distance from the colonies, and Mauritius was no exception. Between the founding of Britain's colonial empire during the seventeenth century and its dismantling during the twentieth century, authority shifted constantly between London and the colonies. The British government set certain norms that colonial elites and political organizations subjected to continual renegotiation. Governance varied from colony to colony as local groups manipulated British models to enhance their own status. Metropolitan influence remained important, particularly in the external relations of the colonies: trade, foreign affairs, and defense. However, the internal affairs of the colonies often remained under the influence of local politicians and organizations.

Colonial administrators retained considerable autonomy from London. Existing technologies for communication and transportation limited London's ability to make decisions for the colonies. As Christopher Bayly argues, the early Colonial Office was politically weak, was understaffed, and spent most of its time contending with the slavery issue. From the 1780s until the 1820s, colonial governors and their staffs imposed a variety of authoritarian measures over their subjects. The wars with the French as well as the occupation of enemy territory provided convenient excuses to imitate the reactionary measures of the British government at home as well as the colonial model of the East India Company. In Mauritius and elsewhere, new British governors inherited authoritarian administrations from their French and Dutch predecessors and in some cases even made them more despotic. There was no question of extending representative institutions to these Crown Colonies. Nevertheless, local groups and institutions influenced even the most despotic colonial regimes, sometimes through rebellion.[47]

The British made only minor changes to the Mauritian political system, but they completely overhauled the island's economy. Had French rule continued, Mauritius might have maintained a diversified economy, with both trade and sugar production. But British rule made it difficult to support a diversified economy, despite any intentions to the contrary. The Navigation Laws prohibited British colonies from trading with foreign merchants. The elimination of free trade caused Port Louis to decline: the only way to survive was to attract British ships *en route* to Asia, but it could hardly compete with Cape Town as a stopover. If Mauritians wanted to make money under British rule, then they had to produce commodities. These had to be sold in British markets to earn the currency needed to purchase manufactured goods and slaves.

The colonists knew from experience that sugar cane could withstand cyclones better than any other crop. Cane-growing had already become somewhat extensive under the French. When the Navigation Laws discouraged foreign trade, the colonists began to invest in land and sugar-milling equipment. But Britain maintained a higher tariff on East Indian sugar than on West Indian sugar. Therefore, Mauritian sugar could not compete on the British market. In 1825, Parliament reduced the East Indian sugar duties, allowing direct competition between Mauritius and the West Indies. The island's sugar industry boomed, and slave labor shifted from the port to the plantations. Between 1816 and 1826, production increased fivefold, from 4,148 to 21,244 metric tons. It continued to climb in subsequent years, cresting over the 100,000 metric ton mark in 1854, and reaching 150,480 in 1862. Between the 1810s and the 1840s, the area cultivated in sugar cane increased from about 4,000 to 25,000 hectares, and by the 1860s there were about 52,000 hectares in cultivation. The number of factories rose from 10 in 1798 to 157 in 1823, and the number kept rising until it reached 303 in 1863.[48]

To meet British demand, Mauritian producers invested considerable amounts of money in improvements to their factories. In 1819 they began to replace vertical cane crushers with horizontal rollers, and in 1822 they introduced steam power to some mills. In 1823, 62 factories were powered by animals, 88 by water, and 7 by steam; by 1848 there were no longer any animal-powered factories, while 4 were powered by wind, 45 by water, and 195 by steam. Factories increased their efficiency in the 1840s when specialized evaporators and vacuum pans for boiling down cane-juice appeared. These enabled the extraction of sucrose from canes to begin to rise from 6 percent to 7.1 percent between the 1820s and 1840s. Technical improvements also allowed factories to expand their capacity, which in turn increased the amount of land they could service.[49]

The owners of sugar plantations tended to seek improvements in the factory rather than in the field, but one estate owner, Charles Telfair, took a strong interest in both natural history and agronomy. He imported the Bellouguet cane from Java and helped revive the Société d'Emulation. In 1829, Telfair established the Royal Society of Arts and Sciences, his most important legacy to the island. He secured the patronage and financial support of the colonial government for the new institution, so that members could offer prizes and conduct research in natural history, medicine, and agriculture.[50] The activities of the Royal Society of Arts and Sciences showed that island intellectuals could make important contributions to agronomy and botany. Their interest in science was one of the factors that helped sugar cane to remain the dominant crop in Mauritius.

In the early years of the sugar boom, local planters provided the knowledge but British investors provided the capital. Resident Franco-Mauritian proprietors constituted the majority of producers, a situation that contrasted sharply with the Caribbean, where absenteeism was the rule. Between 1815 and 1828, sugar prices remained relatively high, and metropolitan investors risked their capital in distant ventures such as Mauritian sugar. Many Franco-Mauritians owned large tracts of land in the countryside and shifted to cane monoculture while pledging most of their crop to British merchants. Commercial houses in Port Louis invested in sugar estates, too, encouraging speculation. So long as the industry remained prosperous, individual estate owners could afford to encumber their properties with up to a dozen separate mortgages. But by 1828, an inevitable glut of sugar on the British market caused prices to begin a downward spiral that shook investor confidence.[51]

The opening of the British market to Mauritian sugar stimulated the local sugar industry, but external investors could be fickle. One of the main things that worried British investors, besides prices, was the supply of labor on the sugar estates. In Britain, advocates of free trade and free labor were attacking Britain's role in the South Atlantic slave trade as well as the Navigation Acts that supported the plantation complex.

When the British conquered Mauritius, slaveowners hoped for a return to business as usual. Britain had already abolished the slave trade in 1807, but under the terms of the 1810 capitulation, they permitted the island to retain its own laws and customs. The advent of sugar cane monoculture changed the nature of island slavery completely. Before the sugar boom, some slaves were agricultural laborers, but mostly they worked as sailors, longshoremen, domestics, and artisans. Beginning in the 1810s, slaveowners transferred, sold, or rented slaves to the sugar estates, where their labor was particularly grueling and hazardous. Today, the descendants of the slaves refer to this period as *létam margoz*, a Kreol expression meaning "the bitter days."[52] Ever since then, this fundamental historical memory has cast a shadow over social and economic relations on the island.

The blockade had produced a labor shortage, and slaveowners persuaded the amenable new governor to ask the Secretary of State for the Colonies to resume the slave trade. The British government was horrified at the idea. Instead of resuming the trade, London ordered the governor to capture slave smugglers and to take a census of the slaves in order to make smuggling more obvious. In 1814 and 1816, the census takers counted about 60,000 slaves on the island, the majority of them men. But between 1811 and 1821, the government estimated that smugglers landed 30,000 slaves while the British patrols intercepted only 19 percent. It is still not known precisely how many slaves entered Mauritius during the British period, but it was probably enough to prevent a net population decline. Slave traders simply adopted riskier new routes and techniques, which had the effect of raising prices and increasing the appeal of other sources of labor.[53]

The imposition of sugar cane's rigorous labor regime came at a time when anti-slavery activists were gaining the upper hand in British politics. In 1823, the House of Commons adopted the principle of "slave amelioration," which required the colonies to prepare slaves for emancipation by easing plantation punishments and by providing religious instruction. Mauritian estate owners gave amelioration as frosty a welcome as is possible in the tropics. In 1826, the Colonial Office ordered an inspection and registration of the entire slave population, but planters devised numerous subterfuges to evade the system. The governors themselves sympathized with the slaveowners, and under their rule amelioration was a dead letter.[54] However, the Colonial Office was subject to pressure from anti-slavery watchdog groups, and in 1828 it created the new official post of Protector of Slaves. This official had the power to investigate and prosecute renegade slave owners, but he ran into planter opposition very quickly. The local courts sympathized openly with the slave owners and obstructed the protector's work.[55]

The Franco-Mauritians who dominated the new sugar industry could protect themselves locally, but now that so much depended on British markets they needed effective lobbying in Parliament, following the example of the West Indian sugar interests. In some ways, the governor did protect the estates against London. The British colonial government in Mauritius had come to rely upon the sugar industry to provide revenues and governors were reluctant to enforce the Colonial Office's policies to "ameliorate" slavery. But the Franco-Mauritian estate owners felt it was best to represent their interests directly to Westminster. In 1831, a local lawyer named Adrien d'Epinay visited London to forestall emancipation. He failed to do this, but he did succeed in obtaining a free press for the island as well as a new constitution. This provided for a Council of Government made up of the governor, seven British officials, and seven representatives of "the chief landed proprietors and principal merchants." The governor still had authoritarian powers under the new constitution, subject to the Secretary of State for the Colonies' veto, and the official representatives would always vote with the governor.[56]

Most importantly, the new constitution sanctioned the participation of the sugar interest in governance. In 1831, when London issued orders tightening requirements for amelioration, the emboldened Franco-Mauritians literally formed a parallel government under d'Epinay, complete with a militia. This convinced the Colonial Office that it would be hopeless to pursue the amelioration policy any further. In 1833, the resistance of Franco-Mauritians to amelioration was one of many factors that persuaded Parliament to emancipate the slaves throughout the empire.[57]

In February 1835, the British colonial government in Mauritius promulgated the Abolition Act, six months after the British West Indies did. Under its terms, Parliament compensated slave owners for losing their human property. It voted £20 million for all colonial slave owners to divide amongst themselves, according to the value of the slaves they lost. The Mauritian share of the pie came to £2,112,632, which compensated for half the value of 66,343 slaves. Large slaveholders obtained a proportionally larger share of the proceeds than small-scale masters, who were not as adept at filing claims.[58] The value of the sugar industry declined, and many estates and commercial houses declared bankruptcy, swallowing up a considerable amount of local and metropolitan capital.

Parliament also compensated slaveowners by requiring slaves to work for their former owners for a fixed amount of time. The Abolition Act stipulated that slaves, although technically free, would have to "apprentice" themselves to their masters for six years in the case of field slaves, and four years in the case of domestics. Parliament intended apprenticeship to be a period during which slaves would learn the responsibilities of free life, but most slave owners interpreted apprenticeship as additional compensation for emancipation in the form of labor. Apprentices bided their remaining time on the estates impatiently. They made very poor laborers, and London pressured the Mauritius Council of Government to end the system. In 1839, the Colonial Office ordered the reluctant governor to proclaim the full emancipation of the slaves. By 1840, virtually all of them had departed the estates. The colonial government, whose interests were now firmly intertwined with the sugar industry, passed numerous ordinances to coerce the freed slaves back onto the estates; the Colonial Office struck them all down.[59]

The British authorities in Mauritius attributed the flight of labor from the estates to the inherent idleness of the slaves, but Richard Allen has shown that there were other reasons for the sugar industry's work force to disappear. In the first place, the freed slaves had powerfully bad memories of work in the cane fields. During the period 1835–1846, the freed slave population also declined, and the ratio of three men to every two women was not enough to ensure population growth. But the most important reason for the departure of the ex-slaves from the estates was that they could find better work in a thriving non-estate agricultural economy.[60]

Emancipation forced the elite to reconsider the methods of sugar production. The estate owners had the option of luring ex-slaves back onto the estates with competitive wages, but instead they decided to obtain cheap indentured labor from India. The colonial government assisted them in this endeavor, not only with the slave compensation payments, but also with tax revenues raised from the ex-apprentices themselves. There is some truth in the notion that a "slave mentality" existed among the Mauritian elite, and that they found it difficult to negotiate a decent wage with potential employees. Higher wages would have also meant changes in field production methods to economize on labor. But during the 1830s and 1840s, the estates were not inclined to experiment because of instability in sugar markets, their high level of indebtedness, and their investment of compensation money into factory improvements. The colonial government was willing to subsidize the importation of cheap indentured labor from India, removing any possible incentive to pay high wages.[61]

Indian Immigration to Mauritius

Rather than pay the high wages to former slaves, the sugar industry and the government recruited cheap labor from India. In fact, before emancipation, producers were already turning to India for labor. The first Indian indentured laborers arrived in 1825, but the government only came to support the idea in 1829. Between 1834 and 1839, the sugar industry used its compensation payments to bring 25,000 indentured Indians to the estates.[62]

Indian immigration to Mauritius raised humanitarian concerns immediately. The first Indian "coolies" who came to the island received no government protection; their contracts with planters determined their conditions of service. In 1837, the East India Company recognized the potential for abuse of indentured workers and passed regulations restricting the "coolie trade." Indenture contracts could only last five years, and now they were made renewable for a subsequent five years. In addition, ships transporting the Indians overseas had to meet basic humanitarian requirements. Nevertheless, word got back to India that indentured laborers on Mauritian sugar estates were living and working in abject conditions. In 1838, the Indian government halted immigration to the colonies while it conducted an investigation that uncovered many abuses. The sugar industry began to starve for labor.[63]

Franco-Mauritian sugar growers applied pressure and British officials backed down. In 1842, the British allowed indentured immigration from India to resume, subject to new regulations. Contracts would be limited to one year, and immigrants would be entitled to a free return passage after five years, their period of "industrial residence." The British colonial government of Mauritius would also appoint and pay for agents in Calcutta, Madras, and Bombay to supervise immigration, under the auspices of the local government. Mauritius would also have to appoint a Protector of Immigrants to supervise labor conditions on the island.[64]

For the next sixty-seven years, the British assisted immigration along these lines. After 1842, indentured immigration boomed; in 1843 alone, 30,218 men and 4,307 women arrived in Port Louis. Representatives of the estates met them at the dockside, and after the 48–hour waiting period required by law, the men signed contracts to work for two and a half piastres (Spanish dollars) per month, plus clothing and rations, while the women had to follow a man to an estate. In 1876, when the Indian rupee replaced the Spanish dollar as the local currency, wages were five rupees per month, with two rupees equal to one piastre.[65] Until the 1860s, immigration continued steadily, with a veritable flood of immigrants reaching the island in the wake of northern India's 1857 rebellion; in 1858 and 1859, 74,343 arrived. In Mauritius, the government and the planters were cooperating to solve the persistent constraint of labor shortage. Immigration to Mauritius decreased only in the late 1860s, when sugar prices began to decline.[66]

Hugh Tinker argued that even after the 1842 reforms, indenture became a "new system of slavery." It is true that indentured laborers replaced the slaves, and the indentures often suffered mistreatment. But recent scholarship is showing that in many ways it was better to be an indentured laborer than a slave. Many Indian immigrants claimed that labor recruiters misled them about the nature of work on the estates, but in India changing social and economic conditions made emigration to Mauritius seem appealing. In addition, although conditions on board the "coolie ships" could be difficult, David Northrup has shown that Indian migrants to Mauritius probably suffered no more than contemporary European migrants who traveled across the Atlantic.[67] And Marina Carter's research has also broken down the comparison to slavery in some other ways. She has shown that migrants often maintained contacts with India and that those who returned home sometimes recruited family members and other laborers to join them in returning to the colonies.[68] Scholars like Carter and Northrup are not arguing that indenture was a "kinder, gentler" form of slavery: they do show that many of its features were brutal. They are arguing that Indian indentured migration deserves to be understood on its own terms.

Even so, Tinker did not draw his comparisons to slavery out of thin air. The British government of Mauritius had little regard for the laborers and continued policies that assisted the estate owners. For example, in 1846 the Council of Government extended the minimum period of indenture to five years, in order to reduce competition for labor. The Colonial Office vetoed the law, but suggested that the colonial government levy a tax upon those laborers who did not sign indenture contracts within their five-year period of industrial residence. London also recommended that indentured laborers pay a tax to defray the cost of their introduction, which the estates had paid previously. Naturally, the Mauritian legislature adopted these suggestions, because most of its members were involved in the sugar industry. In 1849, the Colonial Office allowed three-year contracts, and in 1862, five-year contracts. It stretched industrial residence to ten years, and taxes fell heavily on those laborers who did not agree to a second five-year indenture. The Colonial Office also ensured that the percentage of women immigrants rose from 17 percent in 1844, to 25 percent in 1849, and to 40 percent in 1868, in order to settle the immigrant population in the colonies.[69]

Colonial policy also discriminated against the "old immigrants," former indentured laborers who had served out their period of industrial residency. Beginning in 1851, the police issued passes to them, registering their addresses and occupations. If the police found an Indian outside his home district, or not employed in his stated occupation, he could be sent to prison or to a Port Louis workhouse known as the "vagrant depot."[70] The law also required old immigrants to pay heavy fees to obtain licenses for almost any kind of business, making it difficult to be employed anywhere but on a sugar estate. Regressive colonial taxes also hit the old immigrants hard. For example, light export taxes fell on the sugar industry, while everyone paid heavy taxes on rum and other consumer goods. In turn, the colonial government used its tax revenues to support railroad construction and other projects that favored the sugar industry, rather than to establish social services for the poor people paying the most taxes.[71]

Some estate managers also mistreated indentured laborers. Colonial law allowed them to dock two days' wages for each day of absence, even in cases of illness. The housing for estate workers, known as the "coolie lines," were often squalid and unsanitary. Estate infirmaries were no better, and medical treatment was to be avoided, if possible. The estates often had shops on their premises to cater to the captive worker population, offering credit at usurious rates as a way of binding workers more tightly to the land. During the hard times between 1865 and 1914, many estates also withheld workers' wages. Managers of Mauritian estates, like their colleagues in other sugar colonies, treated their indentured Indian workers at best with paternalism and at worst with unbridled contempt. The legacy of slavery was alive and well in Mauritius, but the process of labor coercion itself had changed dramatically. The sugar industry relied more upon the colonial government to enforce laws to keep ostensibly free labor in the fields, while on the estates themselves coercion now took more subtle forms than the overseer's whip.[72]

The decision to import laborers from India had profound social and political consequences for Mauritius. Between 1834 and 1909, 451,786 Indians arrived in Mauritius to work in the sugar industry, and 294,197 remained on the island. They increased sugar production, and became the majority community on the island, transforming the old three-tiered social structure of Franco-Mauritians, "gens de couleur," and emancipated slaves. The Indians, like the slaves, were a diverse cultural group: 58 percent embarked for Mauritius in Calcutta, while 33 percent came from Madras, and 9 percent from Bombay.[73] Those hailing from Calcutta originated primarily from Bihar and Uttar Pradesh. These included both Hindus and Muslims. Many spoke languages related to Hindustani, such as Bhojpuri, which became one of the lingua francas of the island's rural areas. During the early years, laborers leaving Calcutta also included many people from the hill districts of Chota Nagpur, including Santals, Mundas, and Oraons. A substantial number of Tamils and Telegus arrived in Mauritius from Madras, while a smaller number of Gujeratis and Marathas came from Bombay. Indian merchants also carried on trade with Mauritius, and by the middle of the nineteenth century a small community of Muslim traders from Cutch and Surat existed in Port Louis.[74]

Many Indo-Mauritians preserved their religions and languages and earned a reputation for saving every penny they earned. They also resisted the worst excesses of the plantation regime in the fields and in the courts.[75] During the late nineteenth century, the strength and industriousness of the Indo-Mauritians put them in a good position to take advantage of the numerous changes occurring in the local economy.

During the 1830s and 1840s, a progressive shrinking up of external investment transformed the Mauritian sugar industry in yet another way. Much of the problem related directly to labor disputes. In 1833, when Parliament voted to end slavery in the colonies, investors quite naturally withdrew their support. But in 1834, when the colonial government approved the importation of indentured laborers from India, investors put their money back into Mauritian sugar production. But then in 1839, the suspension of Indian immigration once again placed sugar production on a shaky footing. When indentured immigration resumed in 1843, five London merchant houses invested heavily in the sugar industry, but within a few years, changes in British trade policy caused external investors to lose all interest in Mauritius.[76] During the 1840s, more and more members of the reformed Parliament were taking up the cause of free trade. In 1846, they repealed the Corn Laws, forcing sugar imported from the British colonies to compete on an equal basis with foreign sugar.[77]

From the perspective of British investors, the prospects of open competition seemed not to favor the Mauritian sugar industry. The industry had been investing and borrowing heavily to improve factories. Metropolitan bankers now worried that Mauritian estates might not be able to repay their loans. In 1848, four of the island's five British financial houses collapsed. In the ensuing panic, several sugar estates and the Bank of Mauritius failed as well.

It was true that the free trade legislation of the 1840s undermined the value of Mauritian sugar estates, but the fact of the matter was that between 1843 and 1848 many estates improved their factory equipment and remained fundamentally sound enterprises. Nevertheless, external funds dried up. British financiers, burned in 1832, 1839, and 1848, no longer wished to invest in Mauritian sugar estates, particularly in view of the continuing geographic, linguistic, and legal problems of conducting business there. But local investors recognized the potential of the sugar industry, or at least stood by their friends and relatives who operated the estates. During the 1850s, credit remained tight; but when Mauritians began to sell their sugar in alternate markets in other British colonies, sufficient capital accumulated to generate an increase in production.[78]

The successive labor and financial crises of the 1830s and 1840s were accompanied by a new development in the sugar cane fields: in 1840, the Otaheite cane started to fail. It was being attacked by the bacteria *Xanthomonas vasculorum*, known as "gumming disease" or "gummosis," and it was also suffering from root disease.[79] The diseases, together with labor and investment problems, caused estate owners to consider every possible way to improve production and reduce costs. The Royal Society's agricultural researchers played a significant role in these efforts. In 1845, the society formed a special Comité d'Agriculture, which conducted a survey of agricultural practices on the island's sugar estates. The Comité d'Agriculture collected and published a corpus of knowledge about sugar canes in the society's transactions. It probably helped to popularize the use of Peruvian guano as a fertilizer, which partially offset the effects of cane diseases.[80]

Unfortunately, the Comité was not able to reverse the decline of the Otaheite cane, which insects and bacteria were attacking. This cane was still planted on virtually all lands under sugar cane during the 1840s, but it was suffering increasingly from "gummosis," which reduces factory yields. In 1848, the governor appointed one of Telfair's botanist protégés, a Czech émigré named Wenceslas Bojer, to head a committee to study the problem. The governor also despatched the ship *Elizabeth* to Ceylon to acquire cuttings of new varieties.[81] When the crew unloaded the cuttings in Port Louis, bystanders noticed they were full of spotted caterpillars. The only sensible course of action would have been to destroy the cuttings as soon as possible, but lax security measures made it possible for an uninformed thief to steal them first. Rather than introduce new cane varieties, the government had only succeeded in introducing a new cane pest, the "borer" (*Diatræa venosata*), which remains a nuisance to this day.[82]

The sugar estates blamed the government for the new borer problem, but soon demonstrated that they could do little better themselves. In 1850, a group of planters sent the ship *Reliance* to collect new cane varieties in Java. It returned with a million cuttings of seven different varieties in its hold, five of which were already present on the island; some authorities still believed it was possible to renew a failing cane variety by importing fresh cuttings from abroad.[83] Not surprisingly, they grew as badly as the previously imported varieties. Some entomologists even believe it was the *Reliance*, and not the *Elizabeth*, that introduced the borer.[84] In any case, the estates learned from the *Reliance* fiasco that unless they invested enough money to pay for an expert capable of identifying cane varieties correctly, they would obtain poor results.

Estate owners developed more sophisticated institutions for lobbying the government for better cane varieties and other matters. The Comité d'Agriculture already brought a certain unity to the estate-owning elite, but the government's annual subsidy to the Royal Society made it difficult for the planters to use it as a lobbying organization.[85] In 1853 its president, a lawyer and planter named Gabriel Fropier, drew together his colleagues in the sugar industry to form an independent Société d'Agriculture. This organization would not only pursue the old Comité's objective of improving agricultural practices, but would also represent the sugar industry before the government.[86] During the nineteenth century, the Comité and the Société d'Agriculture shared many of the same members, but it was the Chamber of Agriculture, a core group of the Société's wealthiest and most distinguished members, that rose to prominence.[87]

2

NEW CANES FOR A DYING INDUSTRY, 1853–1893

The Franco-Mauritian elite dominated island politics despite the façade of British rule. Of course, on many occasions the British did have a significant impact upon the Franco-Mauritian estate owners, especially when Parliament emancipated the slaves and repealed the Corn Laws. But these acts only convinced the sugar barons that they needed to lobby the British colonial government of Mauritius even more effectively. The governors were generally receptive, while the Franco-Mauritians continued and refined their lobbying efforts.

Beginning in the 1850s, the sugar industry organized itself to lobby the British colonial government more effectively. This was particularly evident in matters pertaining to sugar cane cultivation. As a result, over the course of the 1860s and 1870s, the Chamber of Agriculture and the government of Mauritius collaborated extensively to procure new sugar canes as well as to create new knowledge about the plant. Scientific institutions evolved in response to a complex mixture of pressures from both local and metropolitan governments, organizations, and individuals. Together, the government and the Chamber built a kind of conventional wisdom about sugar cane plants and agricultural information: research institutions should procure useful canes and produce useful knowledge; and all such canes and knowledge would benefit the island as a whole.

The Chamber of Agriculture came to wield enormous power in island politics because it represented the only substantial industry on the island. The leaders of the Chamber of Agriculture even held many of the "non-official" seats in the Council of Government. The Chamber was particularly vociferous on labor issues, opposing most efforts to ameliorate the condition of estate laborers. It also pressured the government on many other issues related to the sugar industry, especially in marketing, transportation, communications, and forestry. During the latter half of the nineteenth century, a time of intense competition on the world's sugar markets, its lobbying efforts to obtain new cane varieties helped to keep the Mauritian sugar estates alive.

Sugar's Decline

During the 1850s and early 1860s, the sugar industry prospered thanks to favorable labor and market conditions. Indian labor lowered production costs, while the widespread use of Peruvian guano fertilizer increased yields at harvest-time. The introduction of steamships improved access to the outside world, while during the early 1860s the opening of two local railway lines improved inland transportation. In addition, in 1851 Parliament repealed the Navigation Laws that barred foreign ships from British colonial ports. The repeal tripled the volume of Port Louis's trade and increased the amount of capital available to local financiers.[1]

Britain's new trade policies encouraged the Mauritian sugar industry to find new markets. During the latter half of the nineteenth century, parts of India, South Africa, and Australia experienced rapid industrialization and urbanization. These colonies did not yet have significant indigenous sugar industries, while Mauritius was closer to them than any other part of the plantation complex. To transport Mauritian sugar to these markets, the estates sold to the large Indian merchant firms that controlled the Indian Ocean sugar trade. In 1876 Mauritius adopted the Indian silver rupee as its currency in order to simplify transactions.[2]

Despite these new marketing arrangements, the Mauritian economy remained vulnerable. Reliance on undercapitalized local financiers meant that any financial crisis, such as the one that occurred in 1865, would have major repercussions. The island was still subject to peculiar weather, and in 1865 a serious drought hurt production. Mauritius was fortunate to sit astride a major trade route, but any disruption in world shipping could cut it off; in 1869, the opening of the Suez Canal diverted much of Port Louis's trade elsewhere, thereby decreasing the amount of finance capital that local banks could lend to the sugar estates. The island's small size and relatively dense population also made it vulnerable to epidemic disease, and during these years there were several serious cholera outbreaks. The exploitation of laborers also created a powderkeg of social discontent. But starting in the mid-1860s, two persistent problems began to affect the Mauritian economy more seriously than all the other problems combined: endemic malaria at home and a sugar glut abroad.

Prior to the 1860s, Mauritius was free of malaria, but between 1866 and 1868, a terrible outbreak of the "fever" killed approximately 50,000 people, one-seventh of the population. The disease threw production into chaos, and those who could afford it moved from low-lying, mosquito-infested areas such as Port Louis to the higher Plaines Wilhems district on the central plateau. Until the successful eradication campaigns of the 1940s, malaria remained endemic to Mauritius; even today there are still occasional outbreaks. The disease took a particularly heavy toll among the rural working population. Its adverse health effects handicapped the island economy.[3]

The malaria epidemic was an internal disruption to production, but it was the island's dependence on one crop that exposed it to fluctuations in external markets. Unfortunately for Mauritius, between 1868 and 1914 world sugar prices fell steadily. Starting in the 1860s, France, Russia, Germany, Austria-Hungary, and the Netherlands began to flood world markets with heavily subsidized beet sugar. Britain adhered stubbornly to the principles of free trade. As a consequence, market prices for sugar produced in British colonies fell dramatically. Many West Indian cane sugar producers simply could not compete. The only saving grace for the Mauritian sugar industry was the island's proximity to markets in the Indian Ocean basin. The Indian sugar market kept Mauritius afloat, but the sugar estates had to make improvements and raise capital in order to survive.

During the late nineteenth century, the Mauritian sugar industry reduced production costs by centralizing production. Before the crisis, medium-sized, family-owned estates predominated in Mauritius. Economies of scale allowed larger operations to cut costs and procure the most recent technology. During the late nineteenth and early twentieth centuries, joint-stock companies bought land and equipment and consolidated their holdings into estates of more than 400 hectares with factories. This process continues today, so that estates now produce about 60 percent of the island's sugar themselves, and process all producers' canes into sugar. Medium-sized estates between 40 and 200 hectares usually found it uneconomical to process their own canes, although some still exist in the twentieth century, and are referred to as "large planters."

Changes in factory technology influenced centralization to a great extent. In 1880, 80 percent of Mauritian factories were still using the eighteenth-century open-fire batteries of boiling pans that evaporated the water from the cane juice. This method of producing sugar wasted a great deal of energy, compared with factories using vacuum pans. Sugar factories using modern methods of evaporation could burn bagasse, the dry fibrous matter remaining after the sugar cane juice is extracted from the cane, to provide most of their energy needs. Factories with the primitive batteries needed to supplement the bagasse they burned with wood or other fuels. By 1880, Mauritius had reached an advanced state of deforestation, like many other sugar islands, and the cost of wood was high. Savings in fuel made all the difference in an increasingly competitive world market.[4] Economies of scale played an important role in a factory's ability to invest in new machinery, so that the better-equipped factories tended to be larger. In turn, larger and more heavily capitalized factories needed access to larger tracts of cane lands in order to be profitable.[5]

The island's finance system also encouraged centralization. Estate owners borrowed money on a year-to-year basis from local banks to finance their field and factory operations. During the nineteenth century, long-term finance of sugar estates was virtually unknown. Only the proceeds of the crop made it possible for the banks to lend out the same money for the next year's crop. One bad year could upset the entire cycle of finance.[6] Declining sugar prices, scarce capital, disease, droughts, and cyclones buried many sugar estates under mountains of debt. Sometimes estates became the property of their bankers, and the banks did three things to recover their losses. Other times, the banks formed estate holding companies of their own to centralize production. They also sold the land to neighboring estates that were in the process of expanding the capacity of their factories. In other cases, they divided the estates into small plots and sold them to Indo-Mauritians.[7]

Morcellement and the Small Planters

Some sugar estates were turning to their laborers in a desperate scramble for capital. This was because the arrival of so many Indian immigrants had changed the nature of the island's economy. The government and the estate owners collaborated to solve the problem of labor shortage, and in the process managed to create a large population reserve for the estates. Indians faced many hardships in the cane fields. Some resisted the plantation regime, but plenty of others saved their money, completed their indentures, and became petty entrepreneurs within the sugar economy. Much of their success came at the expense of the Creoles, the name now given to the mixed population descended from Africans, Indians, and Europeans.[8] Beginning in the 1860s, the sugar industry was chronically starved for finance capital because of declining sugar prices and remoteness from metropolitan financiers. Estate owners centralized their factories and landholdings, selling off small parcels of land to Indo-Mauritians in order to raise capital, in a process called *morcellement*. Estates had previously sold small amounts of property during the 1830s and 1840s, but *morcellement* lay dormant during the prosperous 1850s. It resumed in earnest with the first signs of trouble in the mid-1860s. The malaria epidemic of 1866–1868 slowed it, but during the 1870s and 1880s, land sales increased. By the turn of the century, about one-third of the cane land was in Indo-Mauritian hands. By 1916, they owned 31,000 hectares, amounting to 37 percent of the land under sugar cane. The plots averaged less than two hectares, and some were no larger than a fraction of a hectare. While they still depended on the estates for credit and cane processing, new classes of small-scale Indo-Mauritian proprietors formed.[9]

Morcellement was a complex process, as Richard Allen has shown in his research on nineteenth-century land transactions. Sometimes owners sold several hundred hectares at a time, then the buyers turned around and sold to other buyers, and so on until the pieces of land were ultimately divided into plots of less than one hectare. Some of the land was marginal, but much of it was cleared and productive.[10] The quality of the land on these small plots has long been a bone of contention between low-yielding small planters and the high-yielding sugar estates, but recent surveys indicate small planters' lands are not inherently inferior. The historical "yield-gap" results more from lack of capital and instruction. The small scale of plots makes irrigating, fertilizing, "derocking," and the use of implements and machines more costly per hectare for small planters than for large estates.[11]

The *grand morcellement* of the late nineteenth century transformed social, economic, and political relations on the island. Indian immigrants and their descendants formed the vast majority of buyers. They may have been the victims of economic and political repression, but their ability to obtain capital for land purchases in difficult times testified to their hard work, frugality, and entrepreneurial skill.

Census returns indicate further "upward mobility" among the Indo-Mauritians, but much of their progress came at the expense of the Creoles. Between 1851 and 1881, Indo-Mauritians moved into all sorts of fields, replacing the Creoles as the island's principal artisans, hawkers, and carters. Indo-Mauritians also bought small plots from Creoles, who lacked the capital to develop the land, and who could not compete with the new Indo-Mauritian small farmers. The classes of Creole small proprietors and artisans that emerged during the era of slavery virtually disappeared, and most of the Creole population moved to the towns and coastal areas. After the 1880s, the Indo-Mauritian community began to reproduce its numbers without immigration. Migration from India declined in importance even though the system remained in place until 1909, with a brief revival in the 1920s.[12]

Historians of Mauritius have usually portrayed the *grand morcellement* as the "rise" of a class of small-scale, family farmers. But there was probably a grey lining in this silver cloud: the estates may have chosen a policy of *morcellement* in order to diminish their risks. During the late nineteenth century, sugar prices were declining. Estates preferred to divest themselves of fixed labor costs. It is entirely possible that they believed that if they could settle Indian laborers and their families on the land, then unpaid family labor would have to absorb the decline in prices; paid laborers on estates were more likely to protest than wives and children on small family plots. Unfortunately, it is not yet known whether this was, indeed, the case. Marina Carter has established the significance of female labor in the sugar economy: even though few women worked as indentured laborers, many earned livings as casual laborers and domestic servants even while growing food crops and raising children. Some women also figured prominently in the peasant economy as labor-recruiters, entrepreneurs, and landholders.[13] But we do not yet have a study like Sara Berry's history of families in western Nigeria, which combines anthropological fieldwork and archival research to show how family structures and other social arrangements influence decisions about natural resource management during periods of economic change.[14]

While it is not entirely clear how family strategies were related to *morcellement*, the division of lands had significant long-term consequences for the estates that owned factories. They could still make a profit from processing small planters' canes into sugar. But far in the future, when times improved for the sugar industry, the estates would have been more productive if they still owned all the land supplying the factories. However, the Indo-Mauritian small planters regarded their land as more than just an asset. The land represented their hard-earned money, security, and prestige. To this day, it is rare for a small planter to sell land to a sugar estate.

The establishment of the Indo-Mauritians on the land also changed the way in which sugar estates recruited their labor. They retained some monthly laborers who still resided in the "camps" on the estate grounds, but now they recruited most of their work force as casual day laborers through the agency of Indo-Mauritian foremen called "sirdars." The typical sugar estate had always selected certain sirdars to act as intermediaries between management and labor. During the early years of indenture, sirdars supervised labor in the fields. Later, estates sent sirdars to recruit bands of laborers in India. Over time, as indentured labor declined in usefulness, sirdars became job-contractors for the sugar industry. Few Indo-Mauritians could earn a living exclusively from their small plots, while the estates still needed extra labor during the harvest. The estates paid sirdars for performing certain tasks with hired labor. The sirdars took charge of the laborers and pocketed a commission from the estate. The estates also refused to buy small amounts of canes from every small planter in the nearby villages; instead, they contracted with sirdars to purchase and collect the canes. In their role as cane brokers, the sirdars advanced money to small planters, often at steep rates of interest. The sirdars represented both sides of the *morcellement* coin; they demonstrated that Indians could prosper in Mauritius, but their existence as a class indicated the continuing power of the large sugar estates.

Some refer to small planters as a monolithic class, but their pattern of landholding suggests two separate overall groups and many different classes. This is probably because the sirdars were among the first to accumulate enough capital to buy their own land. In 1930, the colonial government made its first accurate estimate of small planter landholdings. Defining small planters as cultivators of plots measuring less than forty hectares (100 acres), they found 13,988 plots of less than four hectares in area, and 507 plots of between four and forty hectares.[15] Owners of more than four hectares tended to hire labor, because family labor was usually not sufficient to care for larger plots. The smaller planters used lands under cane to supplement wage income. Their crops provided capital for subsistence as well as investment in land purchases and other enterprises.

Small planters have never been a hopeless, dispossessed rural proletariat. This is despite the fact that they acquired plots because the sugar industry rationalized production, settling its labor force on the land rather than in estate camps. Most small planters found employment off their farms. Although some owned several miniature plots at once, most of them could not support their families based on earnings from less than four hectares of land under cane. They worked as laborers on sugar estates, or as sirdars, job contractors, artisans, and even as schoolteachers, clerks, and professionals.

Family members who could not or would not work for cash outside the household maintained the plot instead.[16] The estates also granted small parcels of cane land to trusted employees, under a system of sharecropping known as *métayage*. In some cases, the estates also rented small planters the "interlines," the area between rows of cane, so they might grow a quick vegetable crop before the canes grew tall and blocked the sunlight. Small plots of land provided extra cash for their owners, in return for a measure of dependency on the estates.

Importing New Canes

At the same time as the sugar estates were selling land to Indo-Mauritians, they were also demanding more help from the government in replacing old sugar cane varieties. The colonial government had already shown some willingness to help the sugar industry by importing cane varieties, but so far it had only succeeded in demonstrating its technical incompetence. Even so, the sugar industry and the government still hoped to farm out sugar cane research to individuals, rather than to fund a research institution. In 1855, the government appointed a committee to study the increasingly serious borer problem. It included two members of the Chamber of Agriculture on the panel, the first evidence of direct cooperation between the sugar industry and the colonial government in scientific matters. The best idea the committee could propose was to offer a prize of £2000 to any person able to stem the tide of the borer "invasion."[17] The committee continued to work through 1856, but made little progress. The members found that burning the cane before harvesting helped kill off some of the insects, but even this did not solve the problem.[18] Through 1858, they received fourteen suggestions; some came from as far away as Java, but none worked. To no avail, the committee voted to extend the prize offer indefinitely and complained that Mauritian sugar cane planters were showing "regrettable indifference."[19] Perhaps these "indifferent" planters sensed the futility of the Borer Committee's efforts, or maybe they hoped sugar prices would rebound enough after 1858's ten-shilling decline to offset any losses.

The sugar industry became increasingly desperate in the face of the borer infestation. In 1860, the Borer Committee advised planters to heat any cane tops used for replanting and to use knives to cut borers out of young canes. These methods were neither effective nor economical, but the committee hoped that if all planters did at least something to kill the borers, then maybe the insects would vanish.[20] They did not. The committee continued to evaluate new methods during 1861 and 1862, including one proposal to import insectivorous birds, and they asked the government to renew the £2000 prize once again.[21]

During the late 1850s, the Chamber advised the estates to replace the Otaheite with other varieties, but the lack of available new varieties undermined their efforts. An 1863 report does indicate that the Chamber's members were beginning to recognize the vulnerability of relying solely upon the Otaheite variety.[22] Precise records of the areas under specific cane varieties were not kept, but by the late 1850s, the Bellouguet (also known as the Diard and by several other names) was becoming popular. This cane had been present on the island for some time already, and it became unhealthy shortly after planters made it their favorite. They replaced it gradually with the Gingham and Bamboo canes during the mid-1860s, but these varieties had existed in Mauritius even longer than the Bellouguet and did not grow well. It is possible that pathogens had already had enough time to adapt themselves to the plants, or that they simply did not thrive in Mauritian conditions, or both.

The Mauritian sugar industry was facing a fundamental biological problem, although at the time cane growers did not realize what was happening. When farmers rely on one single plant such as the Otaheite cane variety over long periods of time, it tends to "deteriorate." Insects and pathogens adapt themselves to the inherited attributes of plants and cause old varieties to decline in usefulness. Occasionally, unwelcome mutations may occur. Growers must substitute new varieties to keep ahead of the declining usefulness of their plants. But new and improved plants are also subject to their own internal mutations and external pathogens, and they, too, must be replaced from time to time. This process of varietal deterioration should not be confused with the vegetative degeneration of the same cane planting from one ratoon to the next. What it does mean, however, is that the more genetic diversity in any given cane-growing region, the better growers can resist varietal deterioration.[23] Relying on a single cane for a long time is dangerous.

During the middle of the nineteenth century, the Chamber grew dissatisfied with the sugar industry's varietal predicament. Some estate owners deduced that it might be a good idea to grow some of the less common canes such as the Penang and the Salangore. Cane growers thought they might be better-suited to their own soils or micro-climates, but neither was truly satisfactory. The Chamber's members clamored for the port authorities to take stricter precautions against introducing new pests and diseases, noting the presence of a new borer, possibly from Madagascar, and a new cane disease in the Flacq district. They also begged growers to be careful when they transferred varieties between the island's different regions.[24] In 1862, the Chamber commissioned a respected local botanist to study the varietal situation. Some members sent away to Spain and Mexico for new varieties, while others favored India and Madagascar. Most of these individual efforts were plagued with technical problems and produced inconsequential results.

The Chamber and the British Botanists

During the early 1860s, the Chamber of Agriculture hit upon the idea of using the government's botanical gardens at Pamplemousses to import cane varieties. The colonial government was slow to accept the idea, but there were three reasons why it was convenient for the state and the Chamber to collaborate: the private efforts of farmers and Royal Society members were producing poor results; the Chamber's cane collectors could benefit from the state's fiscal clout; and the state would derive greater tax revenues from a healthier sugar industry. But the state gardens at Pamplemousses had declined under the ministrations of incompetent British directors, to the point where one visitor enquired of his guide, "But pray, where is the cultivated part?"[25] Between 1849 and 1866, at the behest of British and Mauritian naturalists, a new director improved what one governor termed a "wilderness."[26]

Aside from these significant practical problems, there were other obstacles to using the Royal Botanical Gardens at Pamplemousses, as they were called, for sugar cane research. Theoretically, the gardeners there focused on cultivating a broad range of plants of botanical interest, but they did not grow sugar canes.[27] Still, the Chamber realized that the gardens' staff had a great deal of experience in botany and international plant exchanges. It asked the governor to direct the staff to seek new cane varieties from its international plant exchange partners and to cultivate them in a nursery.[28]

At first, the government responded coolly to the Chamber's proposal for state-industry cooperation in sugar cane research. On 7 May 1862, the governor's colonial secretary wrote to the Chamber that while the governor would ask the director of the Pamplemousses gardens to procure sugar canes in his plant exchanges, the mission of the gardens was to introduce, acclimatize, and propagate other ornamental and economic plants and trees. Therefore, the Chamber could not expect the gardens to do very much to help. The agricultural community should rely instead on its own efforts, and establish experimental nurseries in different parts of the island. In this way, the governor expected the sugar industry to arrive at a "perfect knowledge" of how to cultivate cane varieties.[29]

This response may reveal a classical liberal inclination on the part of state officials not to intervene in business matters, but on a more practical level, the state gardeners at Pamplemousses knew little about cane cultivation. While these discussions were taking place, the gardens' director wrote to the colonial secretary, asking which cane varieties were growing on the island. This way, he could avoid requesting duplicate varieties from abroad. When the colonial secretary forwarded this request to the Chamber, the members were probably struck by the director's ignorance.[30]

Nevertheless, while the senior staff members of the gardens had little experience growing sugar canes, they were still respected scientists within the community of British botanists. The process of staff recruitment for Britain's colonial botanic gardens ensured a relatively high level of practical and theoretical competence.[31] And so, even though this study is concerned principally with Mauritius, it is also necessary to understand the roles of institutions such as Kew Gardens, Britain's central institution for plant research, and the Colonial Office, the central administrative unit for Britain's colonies, in policy debates about Mauritian botany and natural resource management. During the nineteenth century, Kew did not control imperial research outright, but it played an important part in a larger imperial network of botanists.

Kew itself was caught up in political and economic struggles. During the late eighteenth century, the progressive Whig landlords who controlled British politics used Kew to further their interests in colonial trade. Kew helped a network of tropical gardens to arise in the colonies as a way to serve local and imperial interests. The British government intended these gardens to be clearinghouses for economic botany. They were also supposed to be bases for botanizing expeditions. During the early decades of the nineteenth century, Kew and the colonial gardens declined as liberal ideologues questioned state support for scientific research. However, between the 1850s and the 1870s, Kew's administrators revived state interest in botany by linking Kew to "moral imperialism," arguing that a network of colonial gardens would reduce dependency on slave-grown crops and increase the development of "legitimate" agriculture and commerce.[32]

One story that illustrated both the potential and the pitfalls of a central imperial garden was Kew's most famous colonial exploit: the transfer of the cinchona plant. During the 1840s, French military doctors in Algeria proved that quinine, derived from the bark of the cinchona tree, could cure malaria, which was the scourge of Europeans in the tropics.[33] However, cinchona trees grew only in the Andes, making quinine quite expensive. Between 1859 and 1862, the India Office and Kew Gardens sent expeditions to the Andes to steal cinchona trees and seeds. As soon as the plants arrived in England, Kew distributed them to botanic gardens in Mauritius and most other tropical colonies. From 1860 to 1879, all these gardens conducted experiments on the cinchonas, but the results discouraged the establishment of plantations. The Dutch also acquired cinchona from the Andes, but their plantations in Java were much more successful than any plantations in British colonies because they yielded a higher quality drug at a lower cost.[34]

Kew publicized its role in the cinchona transfers in order to promote itself and to highlight the potential importance of colonial gardens.[35] However, Kew's legitimation strategies in London did not necessarily translate into credibility in the colonies. From their perspective, the cinchona scheme could only have shown that Kew did not understand how to apply botany to local agronomy. The fact that in most British colonies the cinchona plantations fared poorly in comparison to the Dutch plantations in Java can only have reaffirmed this interpretation. While Kew's interest in helping itself by helping the colonies could bring benefits to the colonies, on the whole colonial farmers had reason to mistrust the ways in which metropolitan scientists construed their interests.

But Kew was still important. Up until the end of the nineteenth century, at the request of the Colonial Office, Kew Gardens provided all the directors and assistant directors for Pamplemousses. The gardens' staff corresponded regularly with Kew, contributed to Kew's cataloguing of the British Empire's flora, and received several important international botanical journals. But the intellectual connection of the state gardeners to British botany caused some cultural tensions in Mauritius. In the first place, the senior gardeners were all British while most sugar estate owners and staff-members were Franco-Mauritian. Despite the gardens' educational mission, there existed little contact between the staff and various local communities and organizations. The Pamplemousses gardeners could deploy the imprimatur of metropolitan scientific authority to legitimate their position. However, they shared little in common with the sugar-estate owners and managers, who had extensive experience with sugar-cane cultivation. The Pamplemousses gardeners could certainly learn how to grow sugar cane, but the Franco-Mauritians remained somewhat reluctant to entrust these important cane-selection services to people who did not understand the world of the plantation.

During the mid-1860s, local and metropolitan interests debated the research agenda of the Pamplemousses gardens. On the one hand, the sugar industry wanted useful information, while on the other hand British scientists urged "pure" research and teaching. Kew had ambitions to turn Pamplemousses into a significant research center, and in 1865 it appointed a well-known scientist, Dr. Charles Meller, as director of the gardens.[36] Kew also agreed with the governor to retain John Horne, the less-qualified assistant director appointed in 1861, so that he could supervise the work of the gardens' 60–odd laborers while Meller catalogued the flora of Mauritius.[37] In 1865, Kew circulated a memorandum to all the colonial botanic gardens, stipulating that their mission should be primarily to educate the public. Gardeners were to acquire many different plants through extensive local and international exchanges, cultivate them in an orderly fashion, and label them for the public's information. Only a few Mauritians appreciated this approach, however, as the weak sales of director James Duncan's catalogue of the gardens attested.[38] The Mauritians who could read such a book seem to have preferred to focus their agronomical attention on sugar cane rather than on useless plants, no matter how educational. But the scientists at Pamplemousses persisted in the face of public apathy. In 1866, Meller worked along the lines Kew laid out for him, writing that in addition to supervising the gardens' work, he was beginning a museum of economic products, with a herbarium and a library.[39]

The Colonial Gardens, the Chamber, and Cane Introductions

Contrary to Kew's wishes, the Chamber of Agriculture wanted the gardens to be directly involved in sugar cane research. This was not a simple matter of a producer's organization wanting to "capture" a state institution through lobbying. Very gradually, the Chamber's leaders began to recognize that the gardens had great potential as a practical institution. Previously, the planters had acquired sugar canes from the South Pacific, where the production of knowledge and new plants depended directly upon the skill of European voyagers and Pacific islanders. Now, the Chamber turned to Kew-trained botanists to cultivate, interpret, and distribute the sugar canes.

During the mid-1860s, the Chamber started to favor a more systematic, controlled approach to producing new sugar cane varieties. At first, the Chamber attempted to cooperate with the gardens, thinking it could teach the staff some basic things about sugar cane by indirect communications through the governor's office. The Chamber's secretary wrote to the colonial secretary, who in turn relayed instructions to Pamplemousses, describing which varieties were already present on the island and suggesting that the state gardens in Sydney, Australia would be a good place from which to request canes. It remained uncomfortable for the sugar estates and the British botanists to communicate. Mistrust was sufficiently high for the Chamber to remind the Pamplemousses gardeners, via the colonial secretary, to take precautions against new insects or diseases arriving with the imported canes.[40] The Chamber hedged their bets, asking the governor for some of the old Mauritian varieties to be reimported, still believing they could plant new cuttings and thereby renew the old varieties. This is despite the research of one of the Chamber's most prominent members, Louis Bouton, who showed clearly that diseased canes needed to be replaced by new varieties.[41]

While Meller commenced his state-sponsored research on the flora of Mauritius, the sugar industry's difficult situation convinced the Chamber that the island definitely needed new cane varieties. The ecological problems of cane cultivation fueled the Chamber's anxieties. In 1863, a new insect pest called the "pou à poche blanche" (*Pulvinaria iceryi*) began ravaging the cane fields, possibly causing even more destruction than the borer. Although it abated naturally after several years, it reminded sugar planters of their vulnerability. By 1866, virtually all cane varieties had some sort of disease, and the Chamber expressed the hope that new replacement varieties could be found abroad.[42]

It is interesting to note that even while the Chamber and the state were making a collaborative effort to introduce canes, some individual sugar planters were continuing their own independent efforts. Many thought New Caledonia would be a good place to look for new canes. In 1866 one Mauritian made a voyage there, while another corresponded with the island's French governor and obtained nine new cane varieties, including one called Mignonne. Several years later, a Mauritian planter named Louzier made the fortuitous discovery of a mutation of the Mignonne. During the 1870s, the Louzier cane, as it was called, became popular in some parts of the island.[43] It strengthened the case for shifting the cane-hunting from overseas to home. But it was also an indication that sugar estate owners believed that it was worthwhile to collect and select their own canes. Collaboration between the colonial state and the sugar estates would only deepen over the course of the next hundred years, but throughout this period some sugar planters continued their own independent efforts to obtain new canes.

But among most planters, a consensus was emerging: if the sugar industry was to thrive, then it needed to underwrite institutional efforts to discover new canes. Despite occasional frustrations, the Chamber pressed on with its efforts to enlist the assistance of the state gardens. In 1866, the president of the Chamber suggested that members take up subscriptions to finance cane importations, seemingly abandoning hopes of assistance from Pamplemousses.[44] Nevertheless, the Chamber lobbied the governor persistently, requesting the gardens' assistance in acquiring new varieties. The governor finally agreed, and the Council of Government voted £100 for the gardens to obtain foreign cane varieties.[45] Meller began work on the introductions, showing goodwill despite his illness with malaria. He even forwarded a report to the Chamber discussing the advantages of the drought-resistant "China" canes, possibly *Saccharum sinense*, which he once observed growing in Natal. Meller remarked that this cane could be grown to advantage in drier Mauritian locations, even though some earlier attempts to do so had failed. This was the first time a government official ever advised the sugar industry on the subject of cane cultivation. During 1867, Meller also wrote letters to "gardens and societies" in Hawaii and Java, and his contacts promised to send cane cuttings as soon as they could.[46] In 1867, assistant director John Horne wrote to Kew Gardens, asking for assistance in procuring canes from the New World. The sugar industry had not only succeeded in obtaining the use of the state gardens, but now also had the metropole at its service, too.[47]

Cane introductions went ahead full steam during the turmoil of the late 1860s. The malaria epidemic took a heavy toll on the gardens' staff, working as they did in a damp lowland area.[48] In 1868, a major hurricane struck the island, and the debilitated staff spent much of their time simply cleaning up the mess and rebuilding.[49] But they persevered with cane introductions, and in December 1866 the first three new varieties arrived from Java, followed by ten more in April 1867. In August 1868, three varieties arrived from Trinidad, along with a box of unidentified canes from British Guiana, while in October eight varieties came from Queensland. In early 1869, more canes arrived from Jamaica and Penang, and John Horne, now suffering from malaria himself, reported that all the varieties he received grew well.[50]

The Pamplemousses gardens' success in acquiring new cane varieties contrasted with the continued failure of individual efforts. A French admiral who had befriended some Franco-Mauritians during a recent stopover sent a large shipment of canes from Vietnam, all of which arrived dead. Still worse, one grower refused to share any of his reportedly good New Caledonian canes with anyone, even when the Chamber offered him 2,500 piastres for some cuttings.[51] Some Mauritians continued personal efforts to obtain cane varieties. One grower obtained Brazilian canes through the British consul in Buenos Aires and the Société d'Acclimatisation in Paris.[52] But to sort out the problems of cane introductions and to decide on a way to finance them, Dr. Edmond Icery, the president of the Chamber, convened a "cane committee" of five members in July 1868. Icery himself was a saturnine medical doctor and an amateur scientist in the local tradition. He owned two estates in the Flacq district, and during the 1860s and 1870s, he conducted important research in sugar refining and cane cultivation and served six times as the Chamber's annual president.[53] Over time, he worked harder than any others in bridging the gap between British botanists and Franco-Mauritian estate owners.

Icery and his committee established a pattern of industry-state cooperation in sugar cane research that lasted for the next two decades. Reporting back to the Chamber in September 1868, the committee reviewed the history of individual attempts to introduce new cane varieties and bemoaned the failure of the sugar industry to organize a collective cane-importing enterprise. According to the cane committee, members of the Chamber were unwilling to subscribe to such efforts, partly because it was difficult to find a reliable person in other cane-growing regions to collect and ship appropriate canes. Individual planters also mistrusted each other, thinking that their subscriptions to cane-importing would subsidize others who did not subscribe, but who could easily acquire the newly introduced cane cuttings shortly after their arrival. The state gardens at Pamplemousses already had a reliable network for transferring plants from one region to another, and the colonial government could ensure a modicum of fairness if it imposed a tax to support cane introductions. The cane committee's report stated that government intervention would be the only way to ensure the successful acquisition of new cane varieties.[54] Rather than surmount its difficulties through public-spirited organization, the sugar industry turned to the government, in which it had a strong voice.

The cane committee drafted a proposal for procuring new varieties, identifying the best places to find canes and the preferred method of distributing them. It suggested that Meller travel to Hong Kong, Japan, the Phillipines, New Caledonia, New Hebrides, the Society Islands, and Queensland to acquire new canes. The governor supported this state mission and may have even influenced the drafting of the proposal himself.[55] On October 14, 1868, the colonial secretary wrote to Icery that the governor would dispatch Meller to New Caledonia and Australia immediately, and that the Council of Government would place £400 at his disposal. Meller was to acquire as many good canes as possible and to send them back to Pamplemousses. The governor also ordered the state gardeners to take charge of cultivating and distributing the new canes "amongst the different sugar estates throughout the island at such a price as may be sufficient to cover the expense of Dr. Meller's mission, and in the first instance to each estate proportionally according to the crop which it produces."[56] The Chamber finally had a reasonable assurance of acquiring better cane varieties from overseas, and in the bargain the colonial government lost some more of its autonomy.

The history of the relations between the Franco-Mauritian estate owners and the Kew-trained British botanists contradicts the most frequently cited diffusionist interpretation of colonial science, Lucile Brockway's *Science and Colonial Expansion*. Brockway's history of British imperial botany is largely the history of Kew Gardens, Britain's central state agricultural research center, and particularly its role in plant transfers around the British Empire. Brockway argues that Kew's scientists, as part of the imperialist, metropolitan state, directed plant transfers around the globe to ensure the expropriation of plants from the periphery and the concomitant transfer of capital to the metropole. According to her, "Kew, directed and staffed by eminent figures in the British scientific establishment, served as a control center that regulated the flow of botanical information from the metropolis to the colonial satellites, and disseminated information emanating from them."[57]

When Brockway wrote her diffusionist history of Kew in 1979, she had an entirely different political agenda from Pyenson, whose later diffusionist work has a distinctly internalist and neoconservative orientation.[58] By contrast to Pyenson, Brockway drew her diffusionist interpretation from "dependency" theory, which blames Europe and the United States for the persistent underdevelopment of the "Third World." She saw the state, in classic Marxist terms, as a dependent superstructure of capitalism. State research gardens in the metropole like Kew were dependent structures of a dependent superstructure. Local research gardens in the colonies can be considered dependent structures to either the metropolitan garden or to the local colonial state, making them dependent structures thrice removed. "Satellite gardens" on the imperial periphery, according to this view, revolved around the instructions sent from the imperial center of gravity, the Royal Botanic Gardens, Kew.

Brockway broke new ground by relating the globalization of European science and technology to imperialist state structures. A number of scholars have followed her lead, most notably Jack Kloppenburg, who also argues that Europeans extracted important information and technology from non-Europeans in order to fuel their own prosperity and power.[59] However, most scholars who conduct research on colonial politics know that the supposed gravitational dependency of colonial state institutions is full of many anomalies. Under standing administrative and legal procedures, British Crown Colonies like Mauritius were supposed to act as separate states. They supported themselves fiscally and they were linked to the metropole through allegiance to the Crown and its local representative, the Governor. But beyond the realm of formalities, communication and transportation technology during the nineteenth and early twentieth centuries allowed for a great deal of administrative autonomy in European colonies. Colonial governments often got their way in frequent disputes with metropolitan officials and capitalists.[60] Even improvements in imperial telecommunications increased the ability of the colonial "men on the spot" to manipulate central colonial authorities, rather than the reverse.[61]

If imperialist science and technology depended upon colonial institutions, then the picture must have been more complex than Brockway allowed. Her thesis stimulated new avenues for investigation, but her exclusive use of metropolitan archival sources and her reliance upon dependency theory tended to flatten out local, contingent aspects of the globalization story. The dependency approach portrays colonized people as victims, which was often enough the case, but it also diminishes their role in the history of colonial science.

The history of Mauritius shows how colonial science must be understood in a broader context. Evidence from Mauritius tends to support Roy MacLeod's efforts to reconceptualize the history of colonial science. According to MacLeod, science in the colonies was many things to many people: it was a means of enlarging European knowledge; it was part and parcel of colonizing ideologies; it was related closely to the formation of colonial identities; it was a dimension of local, colonial culture; and over time it changed its economic, political, and social relations in tandem with broader changes in imperial organization.[62]

Witnessing Canes at the Pamplemousses Gardens

From the late 1860s until the outreak of the First World War, malaria and declining prices hurt the sugar industry. Low prices forced estate owners to rationalize production and to sell property. This process transformed the island's economy from one in which many many medium-sized factory-estates thrived, to one in which large factory-estates and miniature farms existed in symbiosis. The island underwent a profound cultural transformation, as the Indo-Mauritians became the majority community and sought economic opportunities.

The sugar industry's elite and the British colonial authorities responded by excluding most Indo-Mauritians from colonial politics, just as they had done earlier with the liberated slaves. Creoles and Indo-Mauritians lacked access to the political arena, and they were not allowed to participate in debates about new sugar cane varieties, either. The Chamber and the government assumed that rising new classes of small proprietors would derive prosperity from the same agronomic initiatives that the estates desired. Therefore, small planters would contribute taxes to fund institutions in which they had no voice and from which they derived few immediate benefits. Colonial agronomy did not always favor large farmers over small farmers, as the case of West Africa demonstrates, but in Mauritius the sugar barons pulled the levers of the colonial state.

The Mauritian political situation forced the colonial governors to respond to the vociferous sugar industrialists on the local scene and to the requirements of the imperial government in London. One of the brighter governors, Arthur Gordon (1871–1874), alienated the local community, while another, the brilliant John Pope-Hennessy (1883–1889), fell afoul of the Colonial Office. But these two were exceptional: most governors straddled the fence between local and imperial pressures. Mauritius was never a high priority for the British government, and during the nineteenth century the majority of governors appointed to rule the island were mediocrities. Invariably they stumbled along the path of least resistance, making as many concessions as possible to keep the Chamber at bay while trying not to annoy London lest they jeopardize better appointments in the future. Before and after the constitutional reforms of 1886, the governors often befriended the members of the Chamber whom they appointed to the Council of Government.

The story of the production and distribution of new sugar cane varieties shows that cooperation was less a case of the colonial state securing its position by incorporating the island's main industry, than it was a case of the island's main industry securing its position by incorporating the colonial state. As early as 1862, the Chamber foresaw problems in Mauritian sugar production when it first requested government assistance in procuring sugar cane varieties. The government waited to act until 1868, when the economic situation appeared distinctly pessimistic; but when the government did act, it placed the Pamplemousses gardens at the Chamber's beck and call. The gardens were supposed to serve as the plant research arm

of the colonial government. They were also an important link in an imperial and international network of botanists that took direction from Kew
Gardens in England. Pamplemousses continued to serve as a center of non-
economic research, but between 1868 and 1892, when the Chamber of
Agriculture asked for more cane varieties, the governors made sure that the
gardens' staff made cane acquisitions a top priority.

Despite the convenience of having state garden resources at their disposal, some of the Chamber's members were not altogether thrilled to des-
ignate Meller, Horne, and their assistants as the island's official cane collectors. The Chamber used staff incompetence as an excuse to gain greater
control over the state gardens. Several members questioned the staff's ability to identify and cultivate canes properly. One cautioned that while they
may have been competent botanists "well-versed in the means of flowers
and trees," they might have needed "practical advice in cane culture." Icery
added that he thought it possible to be a good botanist and a poor cane
cultivator at the same time.[63] Icery recognized the gap between British botany
and sugar estate agronomy, and promoted himself as the person who could
bridge it.

The Chamber appointed Icery's cane committee to visit the gardens
periodically as a way of supervising the cane collection and lobbying for
more varieties. The members never seem to have doubted their right to
supervise the work of a state institution and agreed that their cane committee
should meet with the colonial secretary to ask permission to do so.[64] On 18
November 1868, the assistant colonial secretary informed John Horne that
the governor acceded to the Chamber's request, meaning that the gardens'
staff was now required to allow the Chamber's cane committee "to have
access at any time to these canes, and to receive any advice which the committee may offer you on the subject of their cultivation and propagation."[65]

But it seems that the Chamber's visits to Pamplemousses were more
than just an exercise in political dominance. Several historians of early
modern European science have written about the relationship between the
social status of scientists and their epistemological credibility. In his study
of Galileo's relations with his patrons, Mario Biagioli argues that early-
modern Italians "projected contemporary assumptions about social distinction and status on the disciplines, their subject matter, and their methodol-
ogy."[66] Peter Dear has shown that when the Royal Society evaluated reports
on experiments, its members took into account the social status of the
observers. But members also expected descriptions to adhere to rhetorical
norms, no matter how the experiments had actually been conducted.[67]

To be sure, nineteenth-century Mauritius was not early modern Europe. But in some respects, nineteenth-century Mauritius resembled seventeenth-century Europe more than it resembled nineteenth-century Europe: the island had an absolutist government, while local networks of patronage and authority depended upon a landed elite. More importantly, there were significant cultural differences between the Franco-Mauritian estate owners and the Kew-trained British botanists. How were the two groups to establish mutually understood "rhetorical norms" in scientific matters? The history of early modern European science suggests that sugar estate owners might only find the Pamplemousses cane collecting and cultivating experiments believable if members of the Chamber witnessed them and reported favorably upon them. The work of Steven Shapin and Simon Schaffer on witnessing is particularly useful to consider in the Mauritian context. Shapin and Schaffer analyze Robert Boyle's air-pump experiments, which demonstrated that a vacuum could exist. Boyle had to arrange for the right witnesses to report on his experiments in order to achieve credibility.

Even scientists in less diverse colonies depended on these kinds of social and rhetorical practices to gain credibility. Jan Todd has explored the role of persuasion in her book on late nineteenth-century Australia, *Colonial Technology*. Todd borrows liberally from a sociological approach known as "social constructivism." In this view, science and technology never stand alone, but scientists and engineers use complex economic, political, and social networks to promote and support them. These networks depend upon both local and metropolitan support, an insight that raises questions about diffusionist approaches to science, technology, and imperialism. According to Todd, when Australians imported European technologies, they had to build the credibility of these technologies upon local Australian networks of support. By the late nineteenth century, Australia was quite different from Britain and Europe. The descendants of Europeans who lived there were developing a distinct identity, forged in Australian circumstances, just as Mauritians were doing at the same time. How then did the importers of technologies make their products stable and acceptable in the new environment? Todd answers these questions by examining the debates that surrounded the introduction of anthrax vaccine to the Australian pastoral industry and the cyanide process to the Australian gold-mining industry.

Todd pays careful attention to the economic, political, and social struggles that surrounded the local stabilization of these technologies. The French bacteriologists who introduced the first anthrax vaccine had difficulty understanding the social and ecological context of their work in Australia, while local Australian scientists used their understanding of bacteriology to produce an equally effective vaccine. Their understanding of the local context drove Pasteur's scientist-entrepreneurs out of the Australian vaccine business entirely. Todd tells a similar story about the Glasgow metallurgists who tried to license the cyanide process to the Australian gold-mining industry. They depended heavily on middlemen to articulate local and metropolitan understandings of the technology.

The promoters of the anthrax vaccine and the cyanide process also relied heavily upon effective demonstration of their products, a fundamental area of inquiry in the sociology of science and technology. The credibility of new technologies often depends on promoters mustering witnesses to achieve their objectives, particularly when a technology can be associated with economic, social, and political conflict.[68] French and Australian scientists made anthrax vaccine credible first by orchestrating demonstrations to influential pastoralists. The cyanide process not only needed to be demonstrated to a skeptical audience, but judges and expert witnesses had to certify its originality and patentability.

The social constructivist approach calls diffusionist interpretations into question. In Mauritius, as in Australia, promoters of metropolitan knowledge and techniques had to persuade people to adapt them to local economic, political, and social systems. Different classes and cultures came to understand the sugar cane plant in different ways, and influenced the trajectory of research. The technique of arranging for the right witnesses resembles the Chamber of Agriculture's approach to the Pamplemousses Gardens. While any member could visit the gardens and see the experiments for himself, it was more convenient to delegate this authority. Icery's committee became the virtual witnesses of sugar experiments for the Chamber. Witnessing was also a two-way street. The British gardeners could use the methods of virtual witnessing to extend their own credibility to the Franco-Mauritian elite. So long as the Chamber's membership trusted the cane committee, and so long as the cane committee reported favorably upon the cane collecting, then the Pamplemousses gardens had a means to multiply its own credibility throughout the island.[69] The relationship between the Chamber and the gardens had an important bearing upon the interpretation of the new sugar cane varieties that the gardens produced and distributed.

Icery recognized the delicacy of this new relationship between sugar cane producers and the "producers" of sugar cane at the Pamplemousses gardens. Despite the fact that the governor granted the Chamber the power to interfere in the affairs of the gardens, Icery asked his fellow members to restrain their meddling.[70] Nevertheless, the Chamber of Agriculture became deeply involved in the everyday operations of the cane importing project. Meller's canes began to arrive in early 1869, and Icery made the curious statement to his colleagues that the Chamber was putting these plants at the disposal of the government. The reverse was really true, but the distinction between the colonial government and the Chamber was beginning to blur a little when it came to cane importations.[71] Meller continued to send more canes from the Pacific. He wrote to Horne, that after finishing his task of collecting canes in Queensland, Australia's principal cane-growing region, he would be moving along to Sydney and then to New Caledonia. Meller shipped 48 cases of nine different varieties to Horne from Brisbane and hoped to find many more. The cane committee already counted 2,516 plants of five varieties growing at the gardens, and Icery began to worry that the ten hectares devoted to canes at Pamplemousses would soon be filled up. He pressed the Chamber to exercise an active surveillance of these canes and suggested the committee meet with state representatives to create a distribution system.[72]

The Chamber's enthusiastic meddling in the work of the gardens increased as more canes arrived. When some sick-looking canes arrived from Meller in Port Louis harbor, the cane committee joined Horne in inspecting them with a microscope. Even though they had only acquired some mold on the journey, the committee insisted on the pointless half-precaution of planting them in a separate part of the gardens. Horne had no choice but to comply. Icery expressed confidence, however, that some of the canes the gardens had received to date showed promise. On the whole though, Icery and his committee were satisfied with Horne's work in cultivating the new canes. He tapped into the network of colonial botanic gardens and obtained new varieties from Trinidad, British Guiana, Java, Queensland, and other regions. As sugar canes occupied more of his gardens and more of his time, Horne began to worry that he would need more labor and more space to cultivate the canes properly. Not surprisingly, when the colonial secretary asked for the Chamber's assistance, the members seemed willing to help.[73]

As the 1860s wore on, the government lost enthusiasm for the cane-collecting project. Meller never completed his task because he died from malaria in Sydney on February 26, 1869. By then Mauritians were accustomed to the heavy toll the disease was taking, and the Chamber wrote coldly to the governor, asking him "to be good enough to send some other qualified agent, who will proceed to the places originally indicated, and not yet visited by Dr. Meller, in order to complete the mission originally entrusted to him." The Chamber hoped that the Council of Government would vote more money for a new expedition, but the governor proposed sending William James Caldwell, a nearsighted Irish freemason with a some-what checkered colonial career as a teacher, translator, and diplomat.[74] At the time, Caldwell was the Supreme Court's translator, but he claimed, perhaps with a little blarney, that he had enough botanical training to dis-tinguish cane varieties.[75]

Caldwell's appointment indicated that hard economic times were beginning to limit the colonial government's willingness to subsidize plant research.[76] Horne would have been better qualified to go to the South Pa-cific to find more canes, but the Chamber may have felt it more important to keep him attending to the canes already growing in Mauritius, rather than to send him on a dangerous mission. It is also possible that the Cham-ber knew that Horne lacked enthusiasm for the cane-introduction project. He had assumed Meller's duties at the gardens without receiving any addi-tional salary. In 1869, he wrote to Kew that, "I have turned sugar cane planter for the colony, but I question very much any good will come out of it. Exhaustion of the soil from overcropping, and also some others, is the chief cause of what the planters call disease of the cane."[77] Improving culti-vation practices certainly would have helped the sugar industry, but Horne did not yet appreciate what cane planters surmised correctly: that cane va-rieties have a finite economic life span in the face of diseases.

Unfortunately for everyone, Caldwell was not suited to the rigors of cane-hunting. Although his research yielded some useful information, in general he only succeeded in annoying the Chamber. In March 1870, Caldwell's first samples arrived in terrible condition.[78] Rats ate one-third of his second shipment, and to add insult to injury, Icery thought the surviv-ing specimens were identical to canes already on the island.[79] Subsequent shipments contained obvious duplicates, including the Otaheite and Ging-ham canes, as well as many dead canes. The Chamber remained more en-thusiastic than the government about cane collecting and demanded a full accounting of Caldwell's performance.[80]

It can be said in Caldwell's favor that looking for new varieties in an unpleasant place like nineteenth-century New Caledonia was not an easy task. A handful of French settlers and missionaries found him a guide, and he traveled by outrigger canoe along the coast, visiting local cane-growers and collecting their varieties. He experienced all sorts of difficulties: the men he hired to carry the canes liked to chew them, too, and the wardian cases often tipped over his small boat. In Sydney, a longshoreman smashed the glass of one of the cases, and Caldwell had to reglaze it himself. He also could not prevent ships' deckhands from tipping the cases every which way during the journey to Mauritius.[81] The most valuable data to emerge from

Caldwell's journey were his close observations of New Caledonian cane cultivation techniques, but these probably angered the Chamber even more: Caldwell admired the Pacific island "savages," and in his report he argued that their methods for cultivating and selecting canes were superior to many of those practiced in Mauritius.[82] Caldwell's only success was finding and introducing the Wopandou cane to Mauritius, but this went unrecognized. Reintroduced in 1890 as the Striped Tanna, this cane and its mutations, particularly the White Tanna, became the main varieties grown in Mauritius between 1900 and 1930.[83]

Even without the help of useful cane collecting expeditions, Horne's correspondence with cane growers overseas enabled him to acquire even more new varieties. During 1869, two shipments arrived from Rio de Janeiro, reciprocating for canes a Brazilian ship collected in Mauritius fourteen years earlier. The first shipment was damaged and did not grow well, but a second shipment including the important "Uba" variety of *Saccharum sinense*, which Meller had seen growing in Natal, arrived in good condition.[84] To the Chamber's delight, more canes arrived from Java and Hawaii.[85] The cane collection at Pamplemousses grew quite large, and in 1871 the Chamber rented some adjacent land in Pamplemousses so that Horne could expand the gardens' cane nursery.[86] In 1874, Horne reported receiving canes from British Guiana, Trinidad, Jamaica, India, Penang, Queensland, and New Caledonia. The Queensland collection was especially useful, because it included canes from all around the Pacific, particularly Fiji.[87] In 1875, Pamplemousses boasted of a collection of 100 different cane varieties, probably the largest in the world. Many of these grew poorly in Mauritius, and Horne distributed only a limited number of varieties to local planters.[88] Even so, many of these grew well in other parts of the world, and Pamplemousses became an international center for the exchange of sugar cane varieties. Horne also received canes from a number of Mauritians who continued independent efforts to acquire new varieties.[89]

The names that the collectors gave to the new canes reveal a great deal about the entire project of producing and distributing new forms of knowledge about this plant. During the nineteenth century, collectors did not attempt a systematic classification of cane varieties, although by then botanists were well accustomed to the Linnaean system of ordering plants. Great confusion has surrounded cane identification because nineteenth-century cane collectors did not use systematic names. Many canes had different names in different places, such as the Otaheite cane, which was also known as Bourbon and Lahaina. Even so, these names still served a purpose. They highlighted the process of discovery while reminding Europeans of their ascendancy to political domination in the South Pacific and their power on the cane-growing islands themselves. The late nineteenth-century *Annual Reports of the Royal Botanic Garden, Pamplemousses* contain references to canes named after their "discoverers," such as Canne Horne, Diard, and Louzier. Mauritians also named canes after their estates, such as Bois Rouge and Tamarin, which were the sites of reproducing the new plants. Other cane names represented their hopes for success with the new plants, such as Reine, Poudre d'Or, and Avenir. European collectors named some canes after the islands from which they came, such as the Tanna and Fiji canes, while retaining some of the names learned from Pacific island-ers, such as the Tebboe Djoeng-Djoeng imported from Java. European cane collectors and growers felt it was more important to classify canes along spatial lines rather than according to their botanical qualities. The primary objective of their collecting expeditions was to transform what they consid-ered to be nature, plants cultivated by Melanesians, into European productive materials. Their choice of names reflected the sites on the landscape where this process occurred as well as the aspirations of sugar cane planters.[90]

During the late 1860s and early 1870s, the Pamplemousses gardens obtained a full stock of new cane varieties and served as a focal point for selecting and distributing them. The colonial government advanced money to support the acquisition of new canes at the state gardens, while the Cham-ber of Agriculture's cane committee supervised the cultivation and distri-bution of the canes, reimbursing the government with money from their sale. In case the sales receipts did not equal the amount of the loan, the Chamber would support a tax on sugar exports that would affect all grow-ers.[91] In any case, the cane sales proceeded well and the government did not need to levy any new taxes. The Chamber publicized the cane sales through newspaper advertisements, as well as through informal contacts with its members, the proprietors of the island's most important estates. Horne based his selections on sucrose content trials with a nearby factory, although some complained correctly that his methods were not sufficiently rigorous; he crushed all of the canes on the same day, not taking into account the fact that different varieties ripen at different times.[92] In any event, the first cane distribution occurred on 13 April 1869, to be followed by many more before the end of 1891. At Horne's request, the cane committee of the Chamber selected the varieties to distribute and determined their prices as well as the number of cuttings any individual grower could receive.[93] One typical distribution occurred in early 1870. The gardens possessed nearly 29,000 canes of 22 varieties, and the Chamber divided them into batches of 191 assorted canes. Each of approximately 150 estates was entitled to purchase one batch for three pounds sterling, or fifteen piastres.[94] Between April 1869 and December 1875, Horne collected about 20,000 piastres for the canes he distributed. This sum allowed the sugar industry to reimburse the government for the money it advanced for Meller and Caldwell's voyages, plus the costs of cultivating the canes at Pamplemousses, leaving a remainder of approximately seventy piastres.[95]

Horne strengthened his position as director of the Pamplemousses gardens by balancing the interests of local estate owners and Kew Gardens, which was growing more powerful during the 1870s. Kew's scientists were still using the imperial mission to justify their expenditures. Under the direction of William Thistleton-Dyer, the central imperial gardens had some notable successes in promoting colonial agriculture. Dyer corresponded extensively with colonial botanic gardens, advising them on numerous matters. He also influenced the Colonial Office's appointments to the gardens, placing trained botanists in positions where they could conduct research in tropical agronomy and pathology as well as botany. Dyer and his pathologist protégé, Daniel Morris, were instrumental in persuading some colonial governments to do a better job of improving farmers' technical understanding of their crops. Botanic gardens proliferated in Britain's newer colonies, and gardens reemerged from oblivion in the older colonies. Each of these gardens hoped to copy the successes of their older imperial cousins, with encouragement from Kew.[96]

Kew never had direct control over these botanic gardens in the colonies, which could be both a blessing and a curse. Each garden received its funding from the local government of the colony, which controlled its most important decisions. Kew made recommendations to colonial governors on policies pertaining to the gardens as well as on appointments to their staffs. This could prove frustrating when an unsympathetic governor held office, and changes in administration could interrupt the continuity of garden programs. Kew's indirect influence over botanic gardens in the colonies allowed it to take credit for successes but also to distance itself from failures. This was convenient when justifying Kew's budget to the British Parliament.[97]

In his correspondence with Kew, Horne complained that he spent too much time with sugar cane and too little time botanizing.[98] Nevertheless, he had the time to explore the Seychelles, as well as Round Island and Flat Island, both of which lie off the north coast of Mauritius. Following Kew's instructions, he also constructed the herbaceous grounds at Pamplemousses which he used to educate the public in plant classifications. During the early 1870s, an increasing number of people from all classes and communities visited the gardens, perhaps because Horne made it a more interesting place. Kew was impressed, and in 1873 saw to it that Horne was named a member of the Linnaean Society.[99] In the same year, Kew sent out a highly qualified botanist, Nicholas Cantley, to serve as Horne's assistant director.

Despite Kew's importance, the Chamber of Agriculture had come to dominate the gardens' research agenda. The distinction between the colonial government and the Chamber, the island's most powerful producers' organization, was becoming rather vague, at least in the eyes of the producers. Horne used his botanical training and the resources of the gardens to become the island's expert in sugar cane. Some estate owners and managers complained that he was not fair in distributing canes, but for the most part the Chamber was pleased with his efforts. As early as 1871, Icery proposed that the Chamber save any extra money remaining from the cane sales, to create a fund with which to reward Horne's "zeal."[100] The Chamber did not see anything wrong with paying gratuities to government employees, and in 1874 the cane committee recommended paying Horne £100 and Cantley £25.[101] In 1884, while discussing paying another gratuity to Horne and his helpers, some members of the Chamber argued that they had a right to reward Horne because these canes belonged to them, even though they were growing in a state garden.[102] Given the Chamber's level of participation in the management of the gardens and other branches of the government, they had good reason to feel this way. The colonial secretary reminded them, however, that it was illegal for government employees to accept private compensation for their public duties, and Horne never got his money.[103]

Whether or not Horne received a reward, the Mauritian sugar industry was sated with new varieties, and after 1874 cane sales dwindled in popularity. The Chamber sought to wind up the operation and to return the additional land it was renting in Pamplemousses. After some debate, the members asked Horne to continue cultivating each of the new varieties at the gardens. It would be prudent to keep as many as possible, considering how much trouble it took to procure them, even if some canes were of doubtful utility.[104] In 1877, Horne established a nursery for 107 different varieties at Pamplemousses.[105]

Subsequent decisions about the Pamplemousses cane nursery linked the landscapes of state research institutions to the sugar estates, establishing a lasting pattern. The gardens' staff soon realized they needed more land for the cane nursery. At the suggestion of the Chamber, a neighboring estate volunteered 21 hectares for expanding the nursery. The estate's employees also helped the state gardeners in cultivating the canes.[106] Ever since then, some Mauritian sugar estates have benefited from having variety trials on their land; they get early knowledge of good new varieties. Today's cane breeders have also come to depend upon the goodwill of the estates to host their trials, not only because it saves renting additional land, but also because it is only through having field trials in different locations that they can determine which varieties thrive in the island's varying soils and micro-climates.

The Breakdown in Relations between the Chamber and the Gardens

During the late 1870s, cooperation between the Mauritian sugar industry and government researchers continued. In 1877, Horne embarked upon a two-year voyage at Kew's request, visiting the Cape Colony, Ceylon, Australia, Fiji, New Caledonia, and Hawaii. He was mainly interested in other plants, but the colonial secretary asked him to meet the Chamber's demands for new canes. For its part, the Chamber offered to pay shipping costs for the canes, although not for Horne's passage.[107] His first stop in the Pacific was Sydney, followed soon after by Brisbane, where the Queensland Acclimatization Society had an impressive cane collection.

Horne overcame his earlier skepticism about cane introductions and by this time he was a genuine enthusiast. After a short stay in Brisbane, Horne asked the director of the gardens to ship 16 cases of canes to Pamplemousses. Next he made his way to Fiji, Samoa, New Caledonia, and Hawaii, collecting more canes and other plants.[108] During the voyage, according to Cantley, Horne "underwent the greatest bodily fatigue, and . . . imminent danger to his personal safety," which he surmounted only because of his "profound interest in the prosperity of the colony."[109] Horne wrote to the colonial secretary that "The introduction of different varieties of new canes has . . . been of the greatest benefit to the prosperity of the colony. . . . seeing how rapidly many really good sugar yielding varieties of canes deteriorate in Mauritius."[110] Horne acquired 37 new varieties in the Pacific and hoped that ships' captains and missionaries would bring more from regions not yet represented in Pamplemousses' cane collection, namely the Phillipines, Borneo, Celebes, and New Guinea. By that time, he had so many canes that he had to expand the gardens' cane nursery onto land at a government orphanage, three kilometers from the gardens.[111]

After Horne's trip to the Pacific, disease and politics curtailed the role of the state gardens in providing new cane varieties. The entire staff suffered from malaria, "a great hindrance to the work," which caused a high rate of absenteeism.[112] In 1880, Cantley left Mauritius to direct the Singapore botanic gardens. Horne congratulated him on "getting clear of this fever-stricken place." Cantley's replacement as assistant director, William Scott, came down with malaria almost as soon as he arrived in 1881.[113] It was a struggle to keep the gardens functioning.

To make matters worse, Horne's relations with the Chamber of Agriculture began to go sour. His responsibilities as chief government botanist included supervising the forests of the island. When the government passed the forestry ordinance of 1881, requiring farmers to keep their crops at least seventy-six meters (250 feet) away from streams and rivers, it fell to Horne to enforce this unpopular law. Not only did this increase his work without raising his unimpressive annual salary of Rs.5,000, but it made him many enemies in the sugar industry. Although state-industry cooperation continued to be fruitful, there was more bickering during the 1880s than during the 1860s and 1870s. Estate owners from some districts complained that the distribution of canes from Pamplemousses occurred at disadvantageous times for replanting.[114] One owner even resigned from the Chamber in a fit of pique when Cantley and the cane committee forbade him from procuring an inordinately large amount of cane cuttings.[115]

State-industry cooperation in procuring and distributing new cane varieties had never been particularly public-spirited, and during the 1880s and 1890s it became even less so. This was a period of intense *morcellement*, but the Chamber's and the garden's records never mention an Indo-Mauritian purchasing new cane varieties. The large estates controlled the introduction of new varieties, a bulwark against their decline in a depressed sugar economy. But the stress of the times began to show in the Chamber of Agriculture's relations with the state gardens. Fewer estates bothered to pay for trying new varieties, preferring to wait instead for other estates to foot the bill for the cane introductions. As early as 1881, only 25 or 30 estates were purchasing canes at Pamplemousses, and some members of the Chamber chastised their colleagues for being cheap and lacking public spirit. But as one estate owner noted in the same discussion, not buying canes was a rational calculation on the part of the estates. Those buying the canes may have benefited, while those who did not buy them believed they might not benefit and would watch their neighbors' fields instead, just in case they missed a good new variety.[116] On several occasions the cane committee had to allow John Horne to sell the canes at reduced prices, or even to cart them off to a factory just to get rid of them. Horne and Scott kept two or three plants of each variety in a nursery in the event the industry ever needed them, and on the advice of the Chamber they scaled back the cane introductions.

The sugar estates' diminishing enthusiasm for cane introductions was related to the fact that very few of the newly introduced cane varieties had higher yields and better disease resistance than the old canes. In 1891, the Chamber complained that since the late 1860s, only three of the approximately 250 varieties introduced had proven useful. These were the Louzier, which was a mutation of the Mignonne introduced from New Caledonia in 1868; the Port Mackay, introduced from Brisbane in 1869; and to a lesser extent the Bois Rouge, also brought from New Caledonia in 1868.[117] The Chamber was not yet aware that the Tanna cane, introduced from Brisbane in 1890, would soon become the principal cane of the island and would remain in cultivation for thirty years. They also could not know that during the late twentieth century, cane breeders using more sophisticated techniques would be satisfied to find one good cane out of tens of thousands of seedlings. Four useful canes out of 250 was actually quite a respectable result, but from the perspective of 1891, the Chamber felt disappointed. Observing that some of these successful canes already suffered from diseases, the Chamber's secretary-general believed that "after a few years they will submit to the law of degeneration."[118] The Chamber felt it needed to make a stronger effort to obtain new canes. Horne's work had not been sufficient, in the eyes of the Chamber, but he had done much to preserve the sugar economy.

Horne's fall from grace with the sugar industry during the forestry imbroglio made him quite bitter, affecting relations between the government and the private sector in the production and distribution of new sugar cane varieties. The Kew-trained British botanists and the sugar estate owners grew farther apart. While Horne was involved in one forestry dispute, he described a delegation from the Chamber as "a set of ignoramuses who meddle with matters they know not of."[119] Horne complained to Kew constantly in his letters, stating that "Nothing grows here now, intrigues excepted."[120] By 1887, he was writing that, "MISERABLE POLITICS seem to have taken possession of everyone's mind. Some of those lately in power were so crammed full of their own conceit or fads, and all the notions that their friends would stuff into them, that I sometimes got *written to* unpleasantly when I proposed subjects that were not in accordance with their own and their friends' notions."[121]

Horne was understandably cranky. He continued to balance the requirements of the local government and sugar industry on the one hand,

and the metropolitan gardens and Colonial Office on the other, but he was impossibly overworked and underpaid. As a result, the gardens deteriorated. Kew's scientists were pleased that Horne had made some progress in constructing a botanical museum and library, but they were not satisfied with his plant collecting in the Pacific and they worried about the state of the gardens.[122] Horne himself was sick, tired, and lonely. In 1892, he resigned his position and retired. At Kew's urging, Scott replaced Horne as director, but the colonial government reduced the director's salary to Rs.4500. As if things were not bad enough already, the 1892 cyclone devastated the gardens, leaving only a few palms standing. Scott undertook the reconstruction of the gardens along with all his other responsibilities, but in 1897, just as the situation was improving, he died of malaria. No qualified botanist came forward to fill such a poorly paid position, and Kew grudgingly accepted the appointment of a Mauritian nurseryman named Constant van Kiersbilck to direct the gardens. It would be interesting to know about van Kiersbilck's training and affiliations, but the sources are silent.

By 1897, new approaches to the study of the sugar cane plant and the growing power of the Chamber of Agriculture were making the state gardens at Pamplemousses largely irrelevant to the production and distribution of new sugar cane varieties. Nevertheless, the collaboration between the colonial government, the Pamplemousses gardens, and the Chamber of Agriculture in procuring new cane varieties established a restrictive pattern that lasted into the twentieth century, even as the institutional arrangements for producing new sugar cane varieties changed. The state allowed the estate-owning elite to use government facilities to acquire knowledge and new plant technologies, while allowing the elite to limit its distribution to the estates. According to the elite's conventional wisdom, there was never even a question of giving Indo-Mauritian planters direct access to new cane varieties; if the plants proved useful, then Indo-Mauritian planters would obtain them eventually from the estates. The elite gathered and deployed natural resources for their own exclusive purposes. Although Indo-Mauritian planters would not clamor for new sugar cane technology until the 1930s, the seeds of discontent were already planted. The Chamber and the Pamplemousses gardens had worked to establish the credibility of sugar cane research only within the communities of the sugar estates and colonial science. While some new sugar cane varieties grew better than their predecessors, the credibility of the institutions that produced them was only established among a small percentage of the population.

3

CANES IN AN EXPERIMENT STATION, 1893–1913

During the middle of the nineteenth century, colonial botanic gardens co-operated with each other and with local sugar producers to ensure the introduction of new cane varieties. But toward the end of the century, interdisciplinary research stations began to supersede the old worldwide sugar cane network. The production of new sugar canes shifted from the Pacific Islands to the colonial research institutions themselves.

Daniel Headrick argues that this happened because the old system of collecting new plants and transferring them between botanic gardens had reached the point of "diminishing returns." Governments and scientists came to believe it would be best to focus on improving the plants at hand. Intensive breeding efforts required more support than the old botanic gardens could provide. Colonies began to create experiment stations where scientists trained in agronomy, botany, chemistry, entomology, genetics, and mycology studied just one crop.[1]

The creation of new experiment stations in the colonies also reflected the broader shift of the European life sciences from the collecting and classifying work of natural history to the laboratory work of biology. This shift characterized agricultural science as much as it did the other fields of biology. The agricultural experiment station originated in Giessen, Germany, where Justus Liebig formed the first soil chemistry laboratory in the late 1830s. Over the course of the next fifty years, laboratories for agricultural chemistry spread throughout Europe and the United States.[2] Plant selection was also becoming more systematic. For centuries, farmers had been selecting seed from the best plants, and since the eighteenth century some farmers and scientists had also bred hybrids. But by the end of the nineteenth century, some farmers and scientists were trying to make plant selections less empirical and more predictive.[3] However, the principles of plant selection would lack theoretical underpinnings until the rediscovery of Mendel's laws of heredity at the turn of the century.

Even as the new experimental approach succeeded, its spread was not simply a case of new methods and discoveries forcing farmers to reevaluate their approaches to natural resource management. Scientists needed to persuade farmers of the efficacy of new procedures and institutions. They did so through building networks of people, organisms, and inanimate objects to support their claims. Bruno Latour argues that in the late nineteenth-century, Louis Pasteur persuaded the public to believe in his germ theory of disease not only through his research findings but also through the skillful manipulation of politicians and the press. Pasteur did not approach nature and society as two divergent and symmetrical categories, but he intertwined elements of both to construct persuasive arguments.[4] Gerald Geison has gone one step further, showing that Pasteur even resorted to deception in order to manipulate the public's understanding of his scientific practices.[5]

In her book on American agricultural science, Margaret Rossiter shows how it was often difficult for laboratory scientists to persuade farmers to take up the new practices. Rossiter traces how American farmers came to support Liebig-style experimental stations during the middle of the nineteenth century. At first, Liebig's methods found support among university chemists, while many farmers mistrusted the "book learning" of such people. Farmers were not a monolithic group; some supported basic research while some were anti-intellectuals. But they did not simply believe in the new soil chemistry because it was scientifically superior. The new experimental approach became credible partly because it received the support of institutions. During the 1850s, some state legislatures and agricultural societies promoted scientific demonstrations to farmers, a method that continued to be employed by the land-grant colleges and extension services that were founded in the 1860s.[6]

It is not an easy task to import experimental practices into a new setting, which is what Mauritians started to learn in 1892. During that year, Mauritius officially fell in line with the new experimental approach to agriculture. The island joined the network of modern crop experiment stations by creating a Station Agronomique. But the Station Agronomique failed as an institution because its director, while producing competent scientific research, did not produce better canes and did not relate his findings well to the island's farmers. Changes in the production and dissemination of new sugar cane plants can be understood in the context of global changes in colonial research institutions, but they must also be understood in the context of local economic, political, and social relations, even in a remote colony such as Mauritius.

The Political Context for Creating the Station Agronomique

By the time the state and the elite created the Station Agronomique in 1892, the Chamber had gotten even more power to influence the British colonial government. The production of new cane varieties and other agricultural research results became part of the Chamber's larger project of furthering elite dominance of the sugar industry. But during the second half of the nineteenth century, as cane importations proceeded apace, government collaboration with the Chamber of Agriculture raised some thorny issues concerning the state's proper relationship with the society it was supposed to serve. The sugar estates turned to the state to ensure the competent acquisition and distribution of new varieties, while the governors allowed and even encouraged the Chamber to appropriate the services of government officials and institutions. The Chamber may have footed the bill, but in general these activities obscured the distinction between the state and private enterprise. They indicated a new kind of state involvement in the economy, and they also demonstrated the estate-owning elite's ability to browbeat the government. The elite came to believe that it deserved to control the production and distribution of science and technology, and around the turn of the century this attitude hardened.

Initially, the sugar industry's *morcellement* and centralization consolidated the economic and political power of the Franco-Mauritian elite and their assimilated Creole allies. During the late nineteenth century, the elite's grip on Mauritian politics tightened. The governor still had nearly absolute powers, but the changes embodied in the constitution of 1832 gave him the authority to appoint the most prominent Mauritians to the Council of Government, subject to the Colonial Office's approval. Governors generally found that cooperation with the elite was conducive to the smooth running of the colony. But the upper economic strata regarded the Colonial Office, which appointed the governors, as an arbitrary and unsympathetic institution. The Colonial Office could still appoint a difficult and critical governor. Nevertheless, the estate owners spurned new regulations, while frequent administrative changes in Mauritius and in the Colonial Office stalled any momentum for reform.[7]

During the early 1880s, relations between the colonial government and the elite became increasingly tense, particularly on the subject of natural resource management. A major dispute concerning government measures to prevent deforestation pitted the sugar industry against the colonial administration. The first measures to prevent deforestation had taken place under Poivre's administration and continued in force until the advent of British rule.[8] In 1826, forests still covered two-thirds of Mauritius. But the boom in sugar production promoted the clearing of land, as well as the use of wood and charcoal to fire the remaining primitive sugar factories. By 1872, less than one-sixth of Mauritius was covered in forest, and people began to correlate deforestation with soil erosion and periodic droughts. In 1881, the colonial government passed an ordinance making it illegal for anyone to cultivate land within 250 feet of a river. The island is full of small rivers and streams, and the sugar industry felt the government was depriving it of a substantial amount of productive land without compensation.[9]

The "unofficial" members of the Council of Government organized a committee to demand reform of the legislature from the Colonial Office. Their spokesmen argued that the people should be able to elect representatives to decide questions of policy, and they even invoked the memory of the American Revolution. But their definition of "the people" was hardly democratic; they wanted to exclude most Creoles and Indo-Mauritians from the franchise by imposing a high property qualification. The governor saw through these arguments, and he advised the Colonial Office in London to reject their petitions.[10]

By 1883, the stalemate between elite reformists and the colonial government might have led to open revolt had it not been for the arrival of an unusual governor named John Pope Hennessy, a feisty Irishman with rather unorthodox views for a British colonial administrator. In the first place, he was a Roman Catholic, like the Franco-Mauritians and Creoles. But what was even more interesting was that during his previous appointments in the Gold Coast, the Bahamas, the Windward Islands, and Hong Kong, he had advocated greater local participation in governance, to the annoyance of most British officials. He made friends quickly among the Mauritian elite, and in 1886 he was instrumental in persuading the Colonial Office to grant Mauritius a constitution with a partly elective legislature. The new Council of Government included eight "official" British administrators, nine appointed "unofficial" members, and ten elected unofficial members.

Voters had to be males with the ability to pass a literacy test, and they had to have a substantial amount of income or property. These requirements kept the franchise to a bare minimum. By 1909, only 2 percent of the adult population could vote.[11]

Seedling Canes

The sugar barons were tightening their grip on Mauritian politics, wresting a measure of control from the British governors while excluding the vast majority of the population from voting. And while political power was shifting from the governor to the sugar estates, a parallel process was at work in sugar cane research. At the end of the nineteenth century, knowledge of the cane plant was changing dramatically. This new knowledge was due, in part, to the efforts of colonial government botanists. But it was also due to the scientific efforts of plantation owners, peasant farmers, and field workers. In Mauritius, as in other parts of the sugar cane Plantation Complex, the owners of large estates collected observations about cane plants from their colleagues and subordinates in order to reach a new understanding of cane biology. As they did so, they also demanded a greater say in setting the agendas of agricultural research institutions.

As early as 1877, the Chamber's president, Virgile Naz, proposed that the Chamber create an agricultural research center. Naz was a Creole lawyer and landowner who was assimilated into the elite. He also had a strong interest in the sciences, which led him to propose a new institution for research. But the issue of who might have access to the institution complicated and stalled the project. Naz, who was an amateur agronomist, commented at a meeting of the Chamber that not all members had the time or the inclination to devote themselves to experiments on how to raise their crops and select the best new varieties. In addition, estate owners lacked a central place to exchange ideas, and even repeated each others' trials without realizing it. Naz suggested that the Chamber establish a "section agronomique" to keep track of agricultural experimentation on the island.[12]

Some prominent members rejected Naz's idea vehemently. One argued that growers who did not belong to the Chamber would not be paying for the research at Naz's center, and therefore should be excluded from the results, a difficult proposition to undertake. According to him, the general public ought not to have access to privately funded research. Besides, even though a substantial number of Mauritians had an interest in research, a minority of the sugar estate owners and staff members were competent to carry out reliable experiments.[13] Shortly thereafter, the Chamber entertained a proposal from a private nurseryman to grow and distribute new cane varieties. In 1884, the president also suggested that the government subsidize a Chamber scheme to teach agronomy, but nothing came of their efforts.[14] Naz's only accomplishment was to have broached the subject of independent research; other matters, such as price declines and forest regulations, absorbed the sugar industry's attention.

Shortly thereafter, a major new discovery in sugar cane botany recaptured the attention of the sugar industry: sugar cane was found to be fertile, and therefore new sugar canes could be obtained through breeding. But, like some of the scientific discoveries that Thomas Kuhn enumerated in his *Structure of Scientific Revolutions*, the new knowledge of the sugar cane plant did not result from the work of one individual in one discrete moment. Instead, the discovery was the resulted from the cumulative efforts of many different people over a long period of time.

During the nineteenth century, European growers depended on voyager-researchers for the constant provision of new varieties. It would have been preferable to breed new sugar canes themselves, but Europeans misunderstood sugar cane botany so badly that they could not do it. Instead, planters relied on voyagers and state botanists as well as upon their own individual efforts to procure new varieties as the old ones wore out. Research expeditions sailed to the Pacific, and a veritable traffic in Melanesian cane varieties began.

But nineteenth-century cane variety collecting was a decidedly haphazard affair. Scientists could not expect canes that grew well in one region to thrive in another. Cane exchanges could sometimes even be more harmful than helpful, because exchange partners did not observe sufficient precautions to prevent new cane diseases from spreading along with the new canes. Some fumigated the canes to kill surface insect pests, occasionally killing the canes in the process. This procedure did little, however, to harm the bacteria, fungi, and viruses that were the fellow-travelers of the nineteenth-century cane exchanges. For example, in 1882 the botanical garden in Jamaica received forty-four varieties from Mauritius and shortly thereafter sent twelve of these to Barbados. These shipments introduced the fungus *Colletotrichum falcatum* to the West Indies, causing "red rot" disease, a key factor in the decline of the Bourbon (Otaheite) variety.[15]

Even as the Europeans struggled to import new cane varieties, the New Guineans had known all along that sugar cane is fertile. This was because they observed how the plant produces new seedling varieties. But it happens that the Creole and Otaheite canes have the genetic trait of male sterility, and cannot produce seed. European cane growers deduced from this observation that all sugar cane was sterile, despite the evidence before the handful of people who cultivated fertile noble varieties, mainly the Dutch in Java. The New Guineans never passed their full knowledge of the cane to the Europeans, all of whom thought sugar cane required vegetative reproduction.[16]

Over the course of the late nineteenth century, with varietal exchanges at their height, Europeans began to catch on to what the New Guineans had known all along: that canes are fertile and can produce seedlings. In 1858, a Barbadian estate employee noticed some shoots growing in a field of canes. Observing the sharp edges of their leaves, he recognized them as cane seedlings and reported his finding to his employer, who transplanted the seedlings. The plants did grow to be sugar canes, although further experiments showed that the descendants of these particular seedlings were not worth cultivating.[17] Barbadian planters never took full advantage of this new knowledge of the cane plant.[18] Even Kew still refused to recognize the possibility of cane seeds.[19] Gradually the Barbadian planters ceased their investigations.

With male-fertile canes being introduced to replace the Otaheite cane around the world, a number of people in Antigua, Brazil, Java, Martinique, Mauritius, and Réunion also observed cane seedlings in their fields.[20] At first, Mauritians were slow to realize the potential of seedling canes. As early as 1860, an estate manager proposed producing seedlings, but Edmond Icery, the Chamber's president, warned him that if he tried, he would become the laughing stock of the island.[21] In 1871, thirteen years after the discovery of seedling canes in Barbados, another estate owner reported to Icery that he noticed seedlings in one of his fields. Neighboring estate owners verified the claim, but he could not produce a germinated seed as evidence, much to Icery's chagrin.[22] Mauritian naturalists mistakenly confirmed with their correspondents in Réunion and Kew that cane seeds did not exist.[23]

During the same period, cane growers also began to notice and report the phenomenon of variation through mutation, known in sugar-industry parlance as "sporting." In 1868 or 1869, the first definite observation of mutation occurred in Mauritius when a planter noticed yellow sports among striped canes imported from New Caledonia. He separated and propagated the new canes, which became one of the principal cane varieties grown in Mauritius for several decades to come. At the same time, growers in Queensland were also discovering and cultivating different sported varieties.[24]

In Mauritius, Icery continued his observations in spite of the discouragement he received from fellow planters and British botanists. In 1880, he recounted that he noticed some seedling canes himself, but he never could find a remnant of the seed in the roots. He examined a cane arrow under a microscope, observing correctly that out of the tens of thousands of minute flowers, the male flowers were closed. He deduced correctly that this cane was sterile. He added incorrectly that the flower could perhaps be pollinated artificially if one opened up the male flowers. He was wrong.[25] But his efforts did show that colonial planters without metropolitan botanical qualifications could conduct some of the most significant research.

Kuhn argues that while scientific discovery is a cumulative process, it also depends greatly upon individual scientists to observe anomalies that contradict current theoretical understanding of nature. In the world of sugar cane, it was amateur colonial scientists like Icery who, looking at the anomalous data, elaborated a new theory. It is also interesting that the new data about the sugar cane plant came from so many different cultural quarters: from New Guineans, field hands, and plantation owners as well as from botanists and voyagers. Local colonial planters would desire a say in building new research institutions, even as European biology was shifting from the exploration and classification of natural history to the new laboratory research.

The site of sugar cane selection was moving from the gardens of New Guinea to the laboratories and experimental fields of sugar-producing colonies. Planters made some of the most significant contributions to the new understanding of the sugar cane plant, and they came to expect a voice in setting the future research agenda. Therefore, it is not surprising that sugar cane research became caught up in colonial politics.

It was not until 1885, when William Newton assumed the Chamber's presidency, that any serious action took place in the field of improving Mauritian agricultural research. Newton was an aggressive Creole lawyer who had represented the Chamber before the 1874 commission of enquiry. Since then he had acted as the Chamber's spokesman on several occasions. Even after his death in 1915, his friends in the Chamber arranged for him to continue to hector the government; Newton's statue faces Government House in Port Louis, and his right index finger points downward, perpetually inviting the government to grovel. In 1885, he recapitulated the ideas of the Chamber in a pamphlet entitled "The Sugar Crisis." There, Newton complained bitterly about the European bounties on beet sugar, but he nonetheless believed that sugar cane enjoyed natural advantages over beet. Newton suggested that research and teaching could improve the competitiveness of Mauritian cane, adding, "What we want is a few less planters and a few more agriculturists." He proposed the creation of an agronomical research station and training center to achieve this objective.[26]

The Chamber was slow to realize Newton's goal because it proved difficult to negotiate the respective roles of the private and public sectors. Even though the proposed research station was to be private, the Chamber still sought government cooperation, particularly in finance. Governor Pope Hennessy and the Council of Government endorsed Newton's pamphlet and forwarded it to the Colonial Office in London. The Chamber and the colonial government agreed that in order to attract a qualified scientist, they needed to pay the director of a "station agronomique" Rs.10,000, double John Horne's salary at Pamplemousses. The Colonial Office in turn consulted with Kew Gardens, which approved the idea of creating a re-search station, while criticizing how Mauritius relied solely on sugar cane.[27] However, Pope Hennessy became embroiled in disputes with the colonial government, pushing ideas to improve sugar production off the political center stage.

Newton persisted. He used his friendship with Pope Hennessy to press his ideas for improving agricultural science on the island. In 1886, the Colonial Office recalled Pope Hennessy to question him about allega-tions that he meddled in the 1885 elections. Newton accompanied the governor to London as his legal advisor. During the voyage, Newton made the acquaintance of a French agricultural chemist who helped him to sketch out a plan for the station agronomique. His shipmate also recommended that Newton hire Philippe Bonâme, a well-known chemist who was in charge of the agricultural research station in Guadeloupe, to manage the future Mauritian center. Upon arriving in London, Newton met with the Secretary of State for the Colonies and convinced him to approve the cre-ation of a station agronomique.[28]

Newton's meeting with the Secretary of State increased the pressure on the government of Mauritius, but so did recent scientific discoveries in other colonies. While Mauritians bickered, other regions made progress toward breeding canes rather than importing them. Dutch scientists at the East Java Experiment Station (Proefstation Oost Java, or POJ) were pio-neering the science of cane breeding. In 1888, the station's chief scientist published the first professional study of sugar cane's reproductive system and announced the production of seedlings.[29] In 1888, three months after the Dutch published their finding that sugar cane was fertile, experiments in Barbados confirmed the Dutch results.[30] Mauritian planters seemed to discuss the Barbadian results more than the Javan results, possibly because Kew was still acting as a clearinghouse of botanical information the British Empire, and results in Dutch colonies circulated later and less prominently.

The British sugar colony of Barbados was emerging alongside

Mauritius as one of the most significant sites for producing new sugar cane seedlings. Scientists in Barbados were responding to similar pressures, such as cane disease and economic decline. When sugar prices plummeted during the late nineteenth century, Barbadian planters, colonial officials, and Kew's scientists made a joint effort to improve the cane plant.[31] Kew's director, William Thisleton-Dyer, recommended that the Barbadians establish contact with Daniel Morris, director of the state botanical gardens in Jamaica. Morris had just acquired a shipment of 35 new noble cane varieties from Mauritius. Kew and the Colonial Office hoped to disseminate them throughout the British West Indies.[32] In 1884, Morris shipped eighteen of his canes to Barbados, initiating small-scale experiments with varieties. Unbeknownst to Morris, his new canes harbored diseases that would hasten the downfall of the Bourbon variety throughout the West Indies. However, islanders did not recognize the proliferation of new diseases until the 1890s.[33] In 1884, members of the planters' Agricultural Society joined the Island Professor of Chemistry, John B. Harrison, in planting out plots of canes on four estates so that they could make comparisons with Bourbon.

In 1886, the governor insisted that the experiments move to the government's land at Dodds Reformatory, and he hand-picked two scientists to run the experiments. But the governor's decision was problematic in a number of ways. First of all, the land at Dodds was not particularly well-suited to agriculture, although the boys' reformatory provided plenty of cheap labor. Furthermore, the governor named Harrison and John R. Bovell to manage the experiments, a choice that did not inspire confidence, at least not at first. Harrison was a chemist who knew very little about the practical side of planting. Bovell owned a small sugar estate and supervised the reformatory, but he had no training in European botany or chemistry. In time their collaboration would prove fruitful, but initially their results did not impress plantation owners. They conducted trials on the new sugar cane varieties, but their first priority was to study soil chemistry and manures.[34] An article in the planters' *Agricultural Gazette* complained that the Agricultural Society had been excluded from the work, stating that "the Society knows little, and can vouch for little, that goes on at the Station, and this is a great pity, as the testimony of a few (say two) practical planters, of which one should be the president of the Agricultural Society, would certainly not weaken the reports, and may cause them to be more generally believed in."[35] As in Mauritius, the station's credibility depended upon the "testimony" of well-placed witnesses who could report to fellow planters on the research program.

The discovery of sugar cane seedlings at Dodds buoyed the station's local credibility for a short time. Bovell and Harrison employed the same techniques as Mauritian scientists, such as written reports and witnessing, to promote their discoveries. The two researchers presented their seedlings at the Agricultural Society's Exhibition. Harrison reported that large numbers of planters examined them and remarked on their differences from the noble varieties. James Parris, the Barbadian planter who had grown seedling canes during the 1850s, even assisted in verifying the results at Dodds.[36] Bovell opened Dodds to visitors, some of whom came from other parts of the West Indies to verify the seedling discovery.[37] The West India Committee pressed the Colonial Office to grant more funds to the experiments on seedlings, judging their potential to help the entire region.[38] Members of the Barbados Agricultural Society pressed the governor unsuccessfully for money to expand Dodds and to build a second research station.[39]

As time passed the Barbadian scientists could not interest the local elite permanently in seedling canes. This may have been partly the fault of Bovell and Harrison. Their written techniques for publicity were noticeably weak; they buried their exciting discovery in the *Report on the Reformatory and Industrial School for 1888*, which was devoted primarily to the activities of a hundred juvenile delinquents. They also tucked further findings into the back pages of the *Report of the Results Obtained on the Experimental Fields at Dodds Reformatory, 1888*, a booklet that contained mostly statistical reports of fertilizer experiments.[40]

Bovell and Harrison may have failed to reach an audience in Barbados, but in Mauritius, some sugar cane growers took notice of their work. In 1886, the Mauritian colonial secretary's office reported news of Bovell and Harrison's Barbados breeding experiments to the Chamber of Agriculture. The Chamber hesitated to acknowledge the potential of Bovell and Harrison's methods until 1889, when the final results arrived. In fact, Mauritians were only fully persuaded in 1890, when Kew forwarded a package of cane seeds to the governor. The Pamplemousses gardens and several estates tried their hands at germinating them, but only one grower succeeded in obtaining a seedling.[41]

Even so, one seedling was enough to persuade Mauritians that a new era was dawning in their understanding of the sugar cane plant, and it inspired widespread amateur breeding experiments. John Horne asked Kew to "Kindly send me all the papers, your own and those of others, you can get on growing sugar canes from seed. We are all agog on the subject just now."[42] However, he knew from his original failure to germinate the seeds that cane breeding would be a formidable task. Horne predicted that, "Raising cane from seed to get improved varieties will be a long and tedious affair, and there will be many disappointments before a really good, hardy sugar-yielding variety will be obtained."[43]

Founding the Station Agronomique

During the late 1880s, even as sugar cane breeding emerged in other colonies, the idea to establish an interdisciplinary research station remained stalled in Mauritius. In September 1887, the Chamber formed a Station Agronomique Committee to prod the government into action. In May 1888, the committee reported that the island's sugar growers should bear the costs of the Station Agronomique since they benefited from it most directly. The easiest way to raise the funds to provide for the station, according to the committee, was to ask the government to introduce a light tax on sugar exports. The producers would pay to support the station according to the proportion of sugar they produced. The committee estimated that it would cost Rs.60,000 to establish the station and Rs.34,000 to maintain it each year, including Rs.10,000 to pay the director's salary. The starting and annual expenses could be met from a tax of two cents on every fifty kilograms of sugar exported, which was enough to furnish Rs.40,000 each year. The Chamber expected the governor to donate some land at Réduit for the station, reducing its expenses even further.[44]

The proposal seemed straightforward, but there was one especially tricky question: should representatives of private industry or government serve on the board that would oversee the station's work? Not surprisingly, the Chamber's committee recommended themselves for the job, arguing that if the sugar industry was to pay for the station, the Chamber should form a committee to ensure it carried out its mission. They were adamant that the station should not be made a government department.[45]

By then, members of the Chamber were accustomed to having a strong voice in local governance, but they should have known that their request for government assistance in raising revenues would cause the government to demand a voice in setting the research agenda. At the governor's request, the Council of Government appointed its own Agronomic Station Committee to study the question.[46] It included many of the same members of the Chamber who had been working to create a research station for years, including Naz, Newton, and the chairman, Henri Leclézio, who owned an estate in Moka but who was also a rising power in island politics. The presence of several colonial officials ensured that the committee's proceedings were not completely redundant, but in the end their recommendations served the interests of the sugar industry. The committee heard testimony from six factory chemists, a nurseryman, and John Horne of Pamplemousses, allowing for a more informed debate on the mission of the station.[47]

Disagreements about the station's institutional arrangements continued. This caused some planters to take matters into their own hands and explore alternative methods to obtain new canes. Flushed with one estate's success, the Chamber's president made the sanguine suggestion that Mauritius might match the sugar beet breeders, who had doubled the sucrose content of their plant. Following Mauritian standard operating procedures, one owner persuaded the Chamber to name a committee charged with determining and disseminating the best methods for raising canes from seeds.[48]

In 1891, interest in amateur cane breeding peaked, thanks to the Chamber's encouragement, but shortly thereafter it waned when people began to appreciate the difficulties involved. The Chamber offerred a prize of Rs.1,000 to the grower with the best seedling canes and published instructions on how to germinate cane seeds.[49] Perhaps a dozen individuals attempted to grow seedlings, and the award went to Georges Perromat, the manager of Clemencia estate in Flacq, who produced 287.[50] Perromat invested his winnings in a nursery where he specialized in producing seedlings for the sugar industry. In 1893, he sold his collection of 500,000 seedlings to the Mauritius Estates and Assets Company, which planted them at Beau Champ estate in Flacq. Perromat did not take a systematic approach to cane breeding, and by 1894 the company realized that the vast majority of his seedlings were worthless, offerring them to anyone who might be interested. Nevertheless, Perromat continued with his work, and several of his varieties were used as commercial canes, accounting for 20 percent of the area under cane in 1915.[51]

Even as Perromat and other Mauritians bred seedlings, negotiations continued over the creation of the interdisciplinary research station. Finally, the government and the Chamber settled on its research agenda. The sugar industry was coming to recognize that an interdisciplinary scientific approach to agriculture would produce greater profits. Most of the experts testifying before the committee agreed that the station should work on selecting new cane varieties while also addressing soils, manures, plant diseases, and methods of cultivation. John Horne made himself unpopular by arguing for research on crops other than sugar cane, as well as on animal husbandry. Horne also suggested establishing three stations, each one in a different climatic zone, but no one took this sensible suggestion seriously because of his other ideas. In the end, the committee's recommendations for the station's agenda reflected Newton's original ideas, which he himself had derived from correspondence with France's Station Agronomique de l'Est at Nancy.[52]

The new agenda represented a major shift away from the prior work of the Pamplemousses gardens, where the staff worked mainly on collecting and cultivating new varieties. In 1891, the Chamber's secretary spoke of the economic dangers of Mauritius staying "mired in the ideas of yesterday," when agricultural research in Europe and the tropics was becoming increasingly oriented to laboratory experimentation.[53] The Chamber sought to mimic the interdisciplinary laboratory research currently in vogue in Europe. During the same year, the Chamber's president called upon his estate-owning colleagues to "arm" themselves with the "weapons" of science to fight the "common enemy, beet root sugar." He berated the Chamber for taking so long to create a research station, when "the lands are exhausted, the yields diminishing," and the popular Port Mackay cane variety was degenerating.[54]

Throughout the negotiations, the Chamber of Agriculture's representatives on the committee sought direct control over the new station's research agenda. The Chamber proposed "looking for no external aid from government, but relying upon our own exertions," but the government's ability to raise scarce capital through taxation prolonged the collaboration between the state and the sugar industry in scientific matters.[55] While the Chamber accepted some of the Nancy station's suggestions in scientific matters, French recommendations for the management of the institution itself proved distinctly unpalatable. Nancy urged that the Secretary of State appoint a director who would be financially independent of the producers.[56] The Chamber did not consider it "advisable" to make the director dependent on the government, citing their experiences with the director of the Pamplemousses gardens. The Chamber "was of opinion that the station should be placed under their control, inasmuch as they claim to represent the Agricultural Body that would be called upon to pay the cost of the erection and upkeep of the station."[57]

The Chamber claimed the virtual representation of the entire planting community, in much the same way as the Westminster Parliament claimed to represent all of Britain before the reforms of 1832. In other words, they claimed to represent everyone, even though only a tiny portion of the population had voted for them. In Mauritius, the emerging classes of small planters had no voice in government. Had anyone bothered to consult them about the creation of the Station Agronomique, they might have disagreed with the Chamber's position. After all, the Chamber was proposing that all sugar producers, small and large, would pay the export duty to support the station, but only the estate owners who belonged to the Chamber would control the production and distribution of its scientific results. The Chamber even had the audacity to ask the government to give it Rs.5,000 of the new tax revenues to defray its own expenses.

The Chamber's discussions with the government about creating the Station Agronomique, and the work of the committees involved, all produced an entirely unsurprising result: the Chamber got almost everything that it wanted. The export tax would pay for the annual operating expenses and the salaries of the staff. The government forwarded the station Rs.30,000 to construct buildings and Rs.20,000 to equip the laboratories, sums that the export tax would pay back gradually. The station was small, but government backing gave it some potential.[58]

The Chamber also received most of the power to supervise the Station Agronomique. This was an important step, because the production of experimental knowledge is not simply a matter of creating new facts. The credibility of new discoveries rests heavily upon the social and political conventions for regulating and disseminating the new knowledge, as the witnessing of canes at Pamplemousses had already demonstrated.[59] Originally, the Chamber proposed the creation of an oversight board with six members, to be called the Station Agronomique Committee. It recommended that three elected members and one nonofficial appointed member of the Council of Government sit on the committee, together with the presidents of the Chamber of Agriculture and the Royal Society of Arts and Sciences. Most of the individuals holding these offices were likely to be members of the Chamber. The Chamber requested that the committee be able to elect its own chairman as well.[60] The Chamber's oversight committee would witness the Station Agronomique's experiments less directly than the old Sugar Cane Committee that visited the state gardens at Pamplemousses. This may have reflected the more extensive bureaucratic influence the Chamber would have in the new Station Agronomique. In any case, the fact that only members of the Chamber were witnessing the sugar cane experiments did little to make the research credible among people outside of the government and the Chamber.

The British colonial government of Mauritius granted the Chamber most of its wishes, but a struggle took place over whether ultimate control of the station should rest with the government or the Chamber. In February 1891, the Colonial Office instructed the governor to endorse the Station Agronomique Committee's recommendations, but to resist its efforts to appoint and supervise the director.[61] In July 1892 Henri Leclézio, chairman of the Station Agronomique Committee, requested that the Colonial Office reconsider. He argued that committee oversight of the station's director would be important, because planters understood sugar cane agronomy better than government officials. He also argued "that as the sums spent on the station were to be recouped by means of a special tax on sugar it was but just that the planters should have a voice in the management of that establishment." Leclézio proposed that the governor and secretary of state appoint the committee from the ranks of prominent planters. He reassured the Colonial Office that the colonial government's Auditor General would review the station's expenditures because the funds passed through the treasury. The governor endorsed these proposals. He also agreed with the committee's recommendation to appoint Philippe Bonâme, a French chemist completely outside Kew's network of botanists, as the new director.[62] Nobody paused to consider whether the sugar industry's representatives in the Chamber and on the committee might only really represent the large estates with factories.

Diffusionist historians of colonial science like Brockway might lead one to the hypothesis that the Colonial Office and Kew Gardens exercised a monolithic control over the plant sciences in the British Empire. However, the negotiations between metropolitan British and local colonial authorities were sufficiently complex that diffusionism cannot explain them. When it came to deciding who would supervise the new Station Agronomique in Mauritius, the Colonial Office weighed the Mauritian sugar barons' demands favorably against the established pattern of consulting with Kew about botanical institutions. In a minute attached to Leclézio's report and initialled by the Secretary of State, Lord Knutsford, two Colonial Office bureaucrats wrote:

> [1st]: I think we may accede to all these proposals [of Leclézio]; it is so purely a matter of local interest that it is best not to interfere with local wishes, especially as the Director is only to be appointed on a five years agreement. I would not therefore consult Kew as might otherwise be desirable, but assent to what is proposed. . . .
>
> [2nd]: In answering the despatch say that the S of State will not object to the management of the Station Agronomique being placed under the control of a Committee as proposed but ask why the Committee were appointed by the Council of Govt and not by the Governor. . . .[63]

In Mauritius, the Colonial Office's recommendations on the composition of the Station Agronomique Committee were followed by more confusing negotiations. Technically the Colonial Office had the last word, but practically the colonial government of Mauritius could accomplish little without the support of the sugar industry's elite. The Council of Government, including some members of the Chamber, passed a resolution allowing the governor to appoint the committee. The governor allowed the Council to vote on his appointees anyway. The governor also appointed Philippe Bonâme as director of the station in consultation with the committee. The officials of the Colonial Office felt uneasy about the governor's actions, but resigned themselves to a policy of nonintervention.[64] In June 1893, the Station Agronomique opened officially on land adjacent to grounds of the governor's mansion at Réduit.

Philippe Bonâme and the Limits of Private Research

Bonâme accomplished much at the Station Agronomique. He studied all aspects of the sugar industry, but mainly he experimented on different soils and fertilizers to see how they affected sugar production. Bonâme also researched the cultivation of other plants, as well as economics, entomology, pathology, implemental tillage, animal husbandry, and factory chemistry.[65] He established himself as an authority in cane phytopathology after a running dispute over gummosis with Pamplemousses and Kew.[66] In 1901 alone, in addition to analyzing hundreds of sugar cane samples, he investigated irrigation practices, looked into feeding cows a mixture of bagasse and molasses to get them to produce more manure, and weighed the benefits of unleashing imported owls on the island's rat population. A small interdisciplinary research station could not have accomplished more. However, he failed to produce the results that the Mauritian sugar estates desired.

The choice of Philippe Bonâme as director caused some problems, and partly because of him, the station never lived up to its potential. Bonâme was happiest working in his laboratory. Although he was an able scientist, his shyness prevented him from lobbying effectively for the station amongst his loquacious patrons. Throughout the station's twenty-year existence, the Chamber's minutes show that he attended meetings often but spoke rarely. When the members of the Chamber tried to draw him out, as they did in October 1908 when they invited him to a gathering of estate owners and staff members in Flacq, he gave a long, technical discourse on agricultural science that must have bored the amateur audience to tears.[67] Someone with greater social and political skills might have been able to make the sugar industry more aware of the importance of basic research on the sugar cane plant and thereby increase his institution's budget and staff. Bonâme failed to create the kind of political and social network needed to establish a new kind of experimental institution in Mauritius.

Bonâme's problems demonstrate how science, society, and politics are inextricably linked. His efforts came up short in the critical area of providing new cane varieties for the sugar industry. Anyone with a good knowledge of cane cultivation and elementary botany could have participated in the old system of plant transfers, which involved comparing the performance of imported varieties. Bonâme was perfectly capable of this, even though his formal qualifications were in chemistry. Bonâme was also capable of producing and selecting seedling canes, as were many of the amateur scientists among the planting elite. But Bonâme did not do any better than the amateurs in providing new seedling cane varieties suited to the Mauritian sugar industry. Almost every year he used systematic methods to produce more than a thousand seedlings, but none of them were as good as the amateur Perromat's seedlings. In fact, by 1915, five of Perromat's varieties occupied twenty per cent of the area under canes, while Bonâme's canes were nowhere to be seen.[68] Bonâme's slow start in producing seedling canes, coupled with the fact that proper cane selection trials often last for many years, wore the sugar industry's patience thin. Bonâme was trapped between high expectations and modest capabilities.

As early as the 1890s, the Chamber became frustrated with Bonâme's first attempts at seedling cultivation, and they fell back on the Pamplemousses gardens' cane collection to obtain new varieties. The gardens' main success was in distributing the Tanna cane, which became the principal variety cultivated on the island until the mid-1930s.[69] The staff also attempted breeding experiments unsuccessfully and assisted local growers in identifying new seedlings and mutations. Pamplemousses also imported and distributed new seedling varieties from the West Indies.[70] Recognizing the continued importance of the gardens, the Chamber engaged in a long-running dispute with the government over the petty details of cane distributions, which in the end confirmed the Chamber's proprietary rights over the canes.[71] Until 1898, the Chamber continued to use Pamplemousses as a source of canes. In 1901, it used the gardens' collection only as a reserve in case some varieties failed in the Station Agronomique's cooler climate.

The station gradually took charge of producing and distributing new cane varieties for the sugar industry. Bonâme started a repository of living cane varieties at Réduit with the help of the Chamber, which provided cuttings of the fifty-one different varieties remaining at Pamplemousses.[72]

The station also began receiving all the newly imported varieties, at first through Pamplemousses and later directly from foreign exchange partners. During the mid-1890s, Bonâme began scientific breeding experiments, attempting to follow the methods then employed in Barbados and Java. By the early 1900s, he had the sole responsibility of providing canes in Mauritius. The Chamber simply trusted him more than the staff at Pamplemousses, whom one member derided as "well-meaning" but still, after three decades of cane-growing, "a bit foreign to cane culture."[73] They believed in Bonâme because of his superior credentials and also because of the station's institutional ties to the Chamber.

Soon enough, Bonâme himself proved a bit foreign to cane breeding. He used wind pollination to produce seeds, the most common method in the world at that time. Then he selected the best new seedling varieties after several years of field trials. He took the insightful step of planting his seedlings at Pamplemousses as well as Réduit and shared some with estates around the island to see if the plants grew better in particular locales.[74] However, Bonâme's studies were just as flawed as those of the cane breeders in most other regions: his field methods did not allow him to keep track of which parent canes produced better offspring. The consummate laboratory scientist, Bonâme performed extensive chemical analyses on his seedlings, comparing their sucrose content and other qualities with the imported canes. Bonâme's preferences reflected his training as an agricultural chemist. Even though Bonâme was French, giving him some things in common with the Franco-Mauritian sugar estate owners, his intense reliance on European analytical chemistry distanced him from the sugar-growers, who hoped for more useful results. While European chemists were expected to conduct rigorous investigations in the laboratory, Mauritian sugar estates measured the credibility of sugar cane scientists with the yardstick of instrumental reason.

When it came to providing new canes, Bonâme produced little in the way of practical results. In 1907, after a decade of breeding and distributing cane cuttings to the estates, these tests led Bonâme to acknowledge that both in the laboratory and in the field, none of the station's canes performed better than the imported Tannas.[75] He worried that the Tannas might fail like all the previous varieties in widespread cultivation. In 1908, he went so far as to recommend that the Station Agronomique hire a botanist capable of supervising the breeding experiments.[76]

Bonâme's efforts appeared weak when compared with the exciting breeding research in Java. There, scientists were gaining a clearer sense of how to breed canes. The Javan sugar industry was searching for a way to reduce the effects of the devastating "sereh" disease, probably caused by a virus that was then widespread on the island.[77] During the 1890s, a team of Dutch breeders at POJ surmised that it might not be enough only to cross noble canes together to produce better seedlings. POJ attempted hybrid crosses of canes of the noble species, *Saccharum officinarum*, with Indian *S. barberi* canes known to resist sereh. The initial results of these hybrid crosses were disappointing: the new canes did resist sereh, but they were susceptible to the mosaic virus and were not capable of producing sufficiently high sucrose yields to warrant further cultivation.[78] The Dutch scientists turned their efforts back to crossing higher-yielding noble varieties with each other, while planters adopted costly cultivation techniques to reduce sereh. But still, these early hybridizing experiments put POJ on the cutting edge of sugar cane breeding.

The science of sugar cane breeding was evidently still in its infancy during the years of the Station Agronomique's existence. But scientists in other British colonies were also obtaining better results than the station. Generally speaking, their breeding experiments followed the pattern of the work done in Java. Breeding followed empirical selection methods, and crosses of noble canes were producing better seedlings than the hybrid crosses. By 1913 cane growers could choose from at least ten excellent new noble seedling varieties from Barbados, three from British Guiana, and one from Queensland. Barbados even managed to produce a useful hybrid seedling called BH10(12) that grew especially well in high rainfall areas.[79]

The Chamber coveted other cane-growing regions' varieties, and grew dissatisfied with the Station Agronomique's performance. Even Perromat's seedling canes were proving more useful than Bonâme's, and Perromat was a Mauritian nurseryman who lacked scientific credentials. One member of the Chamber reported to his colleagues about the incredibly high yields in Hawaii, not realizing that the islands had a longer growing season than Mauritius, and that they were using the Lahaina cane, the Hawaiian name for Otaheite (Bourbon).[80] Another sang the praises of the Uba cane in Natal, which he claimed could give 25 ratoons, forgetting that this cane contained a great deal of fiber.[81] Others clamored for the introduction of the Badilla cane, highly successful in Queensland.[82] Nobody needed reminding of how breeding had increased the sucrose content of the sugar beet, or of the remarkable experiments in Java and Barbados. Would poor breeding undercut Mauritius in the world's sugar markets?

The sugar industry turned once again to importing new varieties, conceding that the station's breeding program might not produce a useful result for a long time. These imports did not represent a reversion from the restrictive landscape of the laboratory, greenhouse, and nursery to the expansive landscape of the Pacific cane-collecting expeditions. Rather, they simply confirmed the superiority of other cane-producing laboratories in the West Indies and Australia. In 1900, the Chamber acquired seedling canes directly from the Imperial Department of Agriculture of the West Indies.[83] In 1909, Bonâme obtained the Badilla cane and other varieties from the Queensland Department of Agriculture.[84] Between 1904 and 1906, the Chamber even hired a local merchant firm to import new varieties from Barbados and British Guiana.[85] At least one member made independent efforts to find canes abroad, while another may have been raising his own seedlings.[86] Over the course of the next two decades, some of the canes imported from British Guiana were cultivated on a small scale, and were also used as parent canes in breeding. During the decade of the 1900s, none of the newly imported varieties could match the Tannas, according to Bonâme's laboratory analyses.

Importing new cane varieties remained as risky as ever. Some members objected to the Chamber's new policy of cane importations because of the possibility of introducing new diseases; therefore, they proposed strict quarantine measures.[87] The government also feared diseases and required the Director of Forests and Gardens to inspect any imported canes, reducing the Chamber's autonomy to a minor extent. The Chamber still had an arrangement with the Pamplemousses gardens to plant out the imported canes, and the sugar industry soon paid a heavy price for the negligence of government employees. Not only did the plant inspectors fail to detect the introduction of a major new pest, a beetle known as both *Phytalus smithii* and *Clemora smithii*, but the first outbreak of this pest occurred in the vicinity of Pamplemousses, suggesting its connection to government incompetence.[88]

To make matters worse, some members of the Chamber raised questions about the Station Agronomique's utility. As early as 1901, the president wondered why the station was only spending Rs.20,000 each year, when the government was raising annual revenues of Rs.30,000 from the special export tax.[89] Some members said the station could use the extra funds to make itself more useful, while others argued that the government ought to give the Chamber the entire Rs.30,000. One member said that if the station was doing no good, then letting the Chamber do the research directly would save the trouble of running an extra institution. Henri Leclézio, still the chairman of the Station Agronomique Committee, defended the station by accusing his sugar industry colleagues of making unsubstantiated claims. He said that his colleagues in the Chamber never took the trouble to read Bonâme's reports.[90]

Despite the efforts of Leclézio and his supporters, the station's detractors got the upper hand. Leclézio may have been right in accusing the Chamber of not taking full advantage of the station, but the impatient members persisted in trying to undo the institution so soon after its creation. One member argued that Bonâme's chemical analyses were a waste of everyone's time. Any estate could hire a chemist; what the industry needed was more general research on sugar cane. Even Virgile Naz, one of the station's founders, called it a *"corps inert,"* claiming that planters had few tangible results to show for their investment in the institution.[91] Between 1901 and 1906, members proposed using the station's extra money to fund an agent of the Chamber in India, to create a Station Bactériologique, to study how to achieve better telegraphy for the island, and to begin an information service within the Chamber itself.[92] Between 1905 and 1910, the station's committee used the surplus funds to hire a veterinarian.[93] One member even took it upon himself to travel to France to interview a botanist for a position at the station. He argued that although a good botanist would not come cheaply, the industry could ill afford falling behind in cane breeding.[94]

Members of the Chamber criticized Bonâme's experiments for two reasons. On the surface, he produced little in the way of practical results. But he might also have been able to buy himself more time if he had created a network of supporters. Had he worked harder to enlist the support of the Chamber, he might have become more sensitive to their desires and been able to manipulate them more carefully. Bruno Latour gives a number of examples of scientists who persuaded patrons to shift the ways in which they support research. He coins the term "translation" to show how scientists and patrons "enroll" each other into previously separate projects as a way to gain support for them. Of course, the price of this mutual enrollment is that scientists and patrons gain leverage over each other. By Latour's standards, it seems that Bonâme did not have the social or rhetorical skills that would have helped him to enroll the support of the Chamber. Or perhaps he worried that if the Chamber supervised the Station Agronomique too closely, he would not be able to indulge his true love, chemistry.[95]

The Sugar Industry and the Small Planters

During the 1890s and 1900s, as the Station Agronomique rose and fell, the sugar industry weathered a number of crises, but it emerged in a relatively strong position, thanks largely to the close collaboration between the elite and the colonial government. After the devastating cyclone of 1892 and an epizootic in 1902 that killed many draught animals, the colonial government procured substantial British government loans for the sugar industry. The government loaned this money to the sugar estates and invested it in railway construction, boosting the usually capital-starved Mauritian economy. Prices paid to producers remained low, but the area of land under cane increased. During the 1890s, total production had been in the vicinity of 130,000 tons, but in 1903 it crested over the 200,000 mark, and in the good crop years of 1909 and 1913 it almost exceeded 250,000 tons.[96] Even still, the island lacked capital, particularly from outside investors. In 1909, only 13 out of the 141 sugar estates over 40 hectares belonged to foreign companies, all of which were British.[97]

Dramatic changes in the British colonial sugar market also began to have a positive effect on the Mauritian sugar industry. Until the mid-1890s, the British government adhered to a free-trade policy that allowed heavily subsidized beet sugar from Europe to enter Britain. In 1896, an official report convinced the British government that the European beet sugar bounties were killing the West Indian cane sugar industry. In 1902, Britain persuaded the European countries to end their beet subsidies, causing its own colonial cane sugar industries to recover somewhat.[98] This was the first inkling that the British government might once again grant preferential treatment to colonial cane sugar. Nevertheless, Britain still relied primarily on Germany and Austria-Hungary for its sugar. The outbreak of the First World War interrupted this trade, and British sugar prices rose as the fighting destroyed many of Europe's sugar beet farms. The shortage caused cane sugar production to increase all around the tropics, particularly in Cuba and Java. During the war, the British government diverted all exported colonial sugar to its home markets. The empire's efforts at colonial self-sufficiency provided a reliable market for Mauritian sugar, although colonial producers grumbled about price controls, as well as about the new state-supported sugar beet farms in England.[99]

Even as the sugar barons consolidated their formal influence over the local colonial government, other Mauritians began to challenge the status quo. Political discontent began to manifest itself along ethnic lines, although there were also some early signs of interethnic class solidarity. The first rumblings of discontent with elite dominance came mainly from educated Creole professionals and civil servants whose ambitions exceeded their opportunities under the colonial system. The 1886 constitution excluded a majority of them from voting, despite the fact that many considered themselves to be European in culture. They felt they deserved the franchise, which the Creoles of Réunion had already obtained. In 1907, Creole professionals and intellectuals founded a political group called Action Libérale to press for political reform. Despite its bourgeois background, Action Libérale took its case to the streetcorners of Port Louis and the markets of small villages, as well as to the island's newspapers, always a lively forum for debate. Action Libérale united different levels of the Creole community for the first time, and also tapped into the growing discontent of the Indo-Mauritian community. The movement coincided with a government budget deficit and a call from the sugar industry for more government assistance, all of which helped to bring a royal commission of enquiry to the island in 1909. However, the commission did little to alleviate political tensions. Desperate Creoles rioted after Action Libérale's candidates lost the 1911 elections.[100]

At the same time as Creole professionals were entering public debates about the nature of colonial rule, Indo-Mauritians were also becoming more politically conscious. They comprised the majority of the population, and now that many owned land they had a permanent stake in the economy. Their growing awareness of their importance and rights spelled trouble for the old order. During the nineteenth century, Indo-Mauritian resistance to the oppression of the sugar estates typically had not taken any organized form. The estates and villages were comparatively isolated from each other, and many Indo-Mauritians still entertained ideas about returning to India some day. The emergence of more Indo-Mauritian professionals, along with the 1901 visit of Mohandas Gandhi, began to change this way of thinking. The future Mahatma stayed only briefly in Mauritius, but his contacts in the island convinced him that the Indo-Mauritian community lacked self-respect and needed education in its Indian heritage. In 1907, Gandhi sent a young Indian lawyer named Manilal Doctor to teach and defend the Indo-Mauritians. Manilal spent most of his time arguing laborers' cases in court, but he also held meetings in the villages and founded a community newspaper, the *Hindustani*. Before he left in 1911, he agitated for a wider suffrage at the same time as Action Libérale. He was equally disappointed that the 1909 royal commission did not change the constitution.[101]

Even as non-elite Mauritians began to voice their concerns about colonial politics, government officials started to become more aware of the situation of the small planters. Superior attitudes were common among Britain's colonial administrators, but in Mauritius there were some officials who had a more positive evaluation of non-Europeans. In 1885, John Horne noticed the trend toward Indo-Mauritians purchasing cane lands and commented that "The Indian coolie is not only industrious, but he is thrifty, frugal, and saving. . . . This island is sure to become an Indian colony before many generations are over."[102]

As more land passed into the hands of Indo-Mauritians, some government officials wondered whether it might be possible to improve small planter cultivation practices. During the late nineteenth century, Indian boys could receive agricultural education on the estates and at the Pamplemousses gardens, but this was a fairly rudimentary form of instruction.[103] In 1887, the interim governor agreed with Horne that it would "undoubtedly be one step in the right direction if a knowledge of the resources of the colony could be brought home to all sections of the community, and especially to the Indian labourers." He proposed that the government publish leaflets, conduct exhibitions, and offer prizes, but these never materialized.[104] Horne advocated the creation of several agronomical stations to assist small planters as well as estates and sought to enlist the help of the press in spreading agricultural information.[105] Nothing came of these ideas. In 1909 another director of the Pamplemousses gardens was still lamenting the state of agricultural knowledge on the island. He suggested primary school instruction in agriculture as a remedy.[106]

To some extent, the ideas of local officials to ameliorate the farming practices of small planters had their origins in metropolitan thinking. Kew's scientists were developing the notion of making scientific research more accessible to farmers, in what would later be called extension services. During the 1880s, both Kew and Pamplemousses agreed that it was "desirable that detailed cultural and other information of a popular character be prepared for general distribution in the island."[107] Kew's scientists also continued to influence the Colonial Office. The Mauritius Royal Commission of 1909, which approved the establishment of the Department of Agriculture and extension services, was probably familiar with Kew's ideas on the subject.

Lack of instruction limited the improvement of small planter cultivation practices, but their lack of credit was also a serious problem. Any effort to improve small planter cultivation had to address small planter indebtedness. Capital had been scarce in rural Mauritius since the eighteenth century, and the web of rural credit relations caught up all Mauritian farmers. People sought credit at all levels of sugar production, from the largest estate to the smallest sharecropper, to pay for operations and to raise funds for expanding landholdings and factories. The estates depended mainly on local lenders, called *bailleurs de fonds*. As indebtedness increased they often became *de facto* business managers of estates. The estates also obtained capital by selling and renting small parcels of land to Indo-Mauritians.[108] Small planters in turn relied upon the estates for cash and credit to support production. Some borrowed directly from an estate, on the condition that they bring their canes to the estate's factory. Indirect borrowing from the estates was also common. A typical small planter labored on an estate under the supervision of a sirdar, usually a prominent villager owning a plot of more than four hectares. The estates used sirdars as labor contractors; in turn, sirdars employed small planters as estate laborers, taking a percentage of their wages as a commission. In addition, sirdars bought canes from small planters to sell to the factories and also borrowed money from the estates in order to advance credit to small planters at higher rates of interest. The sirdar also arranged for small planters to acquire cane cuttings from the estates for replanting their own fields, and sold their livestock's manure to the estates; cash was so scarce that they preferred to sell manure rather than to apply it to their own fields.[109]

The small planters also obtained credit from small shopkeepers in their villages, from whom they purchased consumer goods. Small planters were chronically short of cash and borrowed money from shopkeepers under a system of rolling debt called *roulement*. By the end of the nineteenth century, most village shopkeepers were Chinese, and they held certain advantages as outsiders. They did not have kinship ties in the village, which helped them in collecting debts. Through their own Chinese kin, shopkeepers had access to a network of finance in Mauritius that extended to the Chinese Chamber of Commerce in Port Louis and possibly as far away as China. The conversion of many Sino-Mauritians to Roman Catholicism may also indicate the tacit support of Franco-Mauritian estate-owners, who reportedly served as godfathers to many Chinese.[110]

Even with these sources of credit, the chronic lack of capital and the constant decline of sugar prices meant that the island economy could not function smoothly. By the decade of the 1900s, the island's credit system was severely overextended. The Chamber of Agriculture applied pressure to the government for a large loan to start an agricultural land bank, but in 1909 the governor and a commission of enquiry pronounced the island's financial system to be essentially sound. This was not to say that there were no serious credit problems. Other colonies managed to attract outside investors, but Mauritius could not because of its exclusive reliance upon a commodity with a declining price. The remoteness of the island from financial centers did not help matters. The continued maintenance of the Napoleonic Code also frightened potential British investors who were not familiar with its impact on business. Between 1892 and 1909, droughts, cyclones, and diseases reduced Mauritian prospects for outside investment even further. Geographical and political conditions made crop diversification and legal reform exceedingly difficult.[111]

The colonial government had largely ignored the credit needs of the small planters, but after a cursory investigation in 1909 it decided that the only way to improve small planter productivity was to remove them from the clutches of rural moneylenders. These people, many of whom were also sirdars and shopkeepers, did lend money to small planters at high rates of interest. But moneylenders were also convenient scapegoats for colonial officials who were unwilling to address larger, more persistent problems in the rural economy. Small planters and large estates suffered equally from declining prices, disease, French inheritance laws, and shortage of capital. The small planters had some special problems that distinguished them as borrowers. They lacked collateral, so moneylenders considered them to be higher risks and charged them higher rates of interest. Sometimes, moneylenders forced small planters to sign over title to their land in advance in case of default. Other things hurt the small planters, too. Crop stealing increased as the island's economy declined, stymying their efforts to grow supplemental food crops for the market. The surra (*Trypanosoma evansi*) epizootic of 1901–1903 also struck them very hard. Many had invested in oxen to haul canes and provide manure, both for themselves and on contract to the estates. When surra killed most of the island's draught animals, the government provided loans to the estates to build their own tramways and small railways. Not only did the small planters lose their animals and their manure, but because of this government railway initiative many also lost their carting businesses.[112]

The small planters' lack of capital had a direct impact upon their cultivation methods, making their cane yields inferior to the estates. Data from 1907 showed that although small planters cultivated about 30 percent of the area planted in cane, they only produced 22.25 percent of the island's weight in canes.[113] This disparity did not result from the supposed inferiority of the small planters' lands, although many people believed this to be true. It resulted from the fact that small planters lacked capital, limiting their ability to replant canes frequently enough, or to purchase enough fertilizer, or to hire enough labor.[114] The dependence of the small planters on the estates with factories for credit and the processing of canes also reduced their yields. Factories retained all the by-products of cane processing, including molasses and "scums," which were used as fertilizers. The small planter's loss in potential fertilizer was the estate's gain, and small planters received no additional compensation for it.

The web of rural credit limited the ability of small planters and villagers to challenge the *status quo*. Even the handful of Indo-Mauritians
who were wealthy enough to vote for representatives in the Council of
Government were vulnerable to intimidation by elite Franco-Mauritian
creditors.[115] When small planters testified before the 1909 commission,
they could not discuss their situations openly and supported the estate
owners' idea of requesting government loans.[116] In the words of a political
activist of the time, the small planters were "in the palms of the factory
owners," and if the commission had interrogated them behind closed doors,
then they would have gotten more forthcoming responses.[117] Nevertheless,
the commission suggested that the proposed Department of Agriculture
address the problems of the small planters through the institution of extension services and cooperative credit societies. Given the prior influence of
the Mauritian elite over the colonial government, the commission might
have anticipated that the sugar barons would shape the new department to
their own advantage. The new department was to have a minimal impact
on the small planters.

4

THE DEPARTMENT OF AGRICULTURE, 1913–1930

The Chamber of Agriculture hoped to broaden the mission of the Station Agronomique while increasing planter control over research. Paradoxically, the Chamber enlisted government support in doing so. As one member put it, ideally the sugar industry would obtain a government Department of Agriculture like the West Indies had, composed of a number of specialists in entomology and phytopathology, as well as chemistry and botany. Another member thought that such a government department ought to be placed under the direct control of the Chamber through the Station Agronomique Committee.[1] This was a totally unrealistic expectation to have of a colonial government's department, but the Mauritian elite's deep involvement in local governance fostered such hopes.

The movement to form a government department of agriculture was not unique to Mauritius. During the first decades of the twentieth century, nearly every substantial British colony acquired a new agriculture department. Typically these were formed from the kernels of the old botanic gardens, but they followed the trend in biology away from global classification and botanical transfers and toward laboratory research into local agricultural and ecological problems.

The new colonial departments also represented a decentralization in the organization of imperial agricultural research. Both Kew and the Colonial Office wanted to create colonial departments in order to make imperial agriculture more competitive. But as Richard Drayton has shown, by

1914 the departments had the unintended effect of reducing Kew to a less significant role in imperial science. Kew promoted colonial laboratory science and agriculture departments, but paradoxically this led to a decline in its own importance.[2] After Kew's apex at the turn of the century, no other imperial institution replaced it. The British government did make one early effort to have the Imperial Institute take over colonial economic advising and the analysis of new products, but the Institute was not suited to the task. The largest colonies and dominions were having greater success conducting their own research.[3]

The Example of the British West Indies

Between 1906 and 1909, the Mauritius Chamber of Agriculture entered into negotiations with the colonial government on the subject of changing the institutional arrangements for agricultural science. The Chamber continued to propose creating a Department of Agriculture, citing correspondence with the Imperial Department of Agriculture in the West Indies, headquartered in Barbados. The government of Mauritius agreed in principle, and even suggested creating an agricultural college to go along with it, again following the West Indian model.[4]

Mauritian sugar barons and British colonial officials looked to the West Indies for examples of how state-supported scientific research could help sugar industries, but the Imperial Department served as more than just a case in point. The founding director of the department in the West Indies, Daniel Morris, was responsible for training the two men who would become the first directors of agriculture in Mauritius, Frank Stockdale (1913–1916) and Harold Tempany (1917–1929). Therefore, the early work of the Imperial Department of Agriculture in the West Indies had a direct bearing on the evolution of institutions and research in Mauritius.

During the late nineteenth century, Barbados was the most important center of sugar cane research in the British West Indies. Bovell and Harrison's experiments with seedling canes were at the forefront of world sugar cane research during the 1880s. But even with their scientific and social credentials, and even with their early successes, Bovell and Harrison still had to continue to produce tangible, practical results. Unfortunately for them, the original seedling canes turned out to yield poorly. By the early 1890s, Barbadian planters were losing interest in the field experiments.

The scientists at Kew also felt ambivalent about Bovell and Harrison. After all, Bovell was a planter with no recognized scientific training, and

Harrison could be a difficult person. During the late 1880s, Harrison became embroiled in a long-running dispute with Daniel Morris of Kew about the discovery of seedling canes. Morris questioned the accuracy of Harrison's research, while Harrison accused Morris of trying to take credit for the discovery. Their dispute spilled over into the British and West Indian newspapers.[5] Nevertheless, however skeptical Morris may have been about the professionalism of science in Barbados, he made sure that Kew still supported research there. This was important because Barbadian planters were losing confidence.[6]

In the early sugar cane research institutions in Barbados and Mauritius, sugar estate owners sought to unite tacit knowledge of local agriculture with globally recognized scientific methods. The Barbadian planters had succeeded initially in combining Harrison's scientific training with Bovell's practical knowledge. But Harrison departed for British Guiana in 1890, so that during the 1890s the Barbadian planters faced the opposite predicament then faced by their Mauritian counterparts. Bovell managed the Dodds station without any globally recognized scientific qualifications, while Bonâme ran his Station Agronomique without a ready sense of local farmers' needs. The crises in confidence that resulted among the large-scale sugar growers of the islands provided rationales for the creation of state departments of agriculture.[7]

In 1897, with the Bourbon cane failing and sugar prices falling, a royal commission toured the British West Indies to assess the economic situation in the islands. In view of the commission's recommendations, Parliament voted grants to the islands to improve communications and agriculture. The British government also created an Imperial Department of Agriculture for the West Indies, with a headquarters in Barbados that coordinated the activities of subsidiary state agricultural agencies thorughout the West Indian colonies. The Secretary of State appointed Daniel Morris as the director of the Imperial Department, arguing before Parliament that Morris held strong metropolitan scientific credentials and that he also had practical experience in the West Indies.[8]

Morris started the Imperial Department quickly, demonstrating his flair for organization. Most of the scientists whom he recruited held metropolitan credentials, which were highly sought after among the Caribbean elites. He used his impressive reports and his ties with Kew to maintain good relations with members of the British parliament, who voted most of the department's funds. However, Morris did allow personal disputes with scientists to influence his decisions for organizing research. Before the creation of the Imperial Department of Agriculture, significant sugar cane breeding experiments had taken place in Trinidad and British Guiana as well as Barbados. Experimental plots in diverse locations could only help the department's research program. But Morris was not able to resolve disputes that emerged between the directors of the old botanic gardens and the new local departments of agriculture in both colonies. To make matters worse, Morris continued his personal feud with Harrison, who was in charge of sugar cane research in British Guiana.[9] Morris decided to centralize the most important sugar cane breeding and agronomical experiments in Barbados under Bovell and himself, with the help of J.P. d'Albuquerque, Harrison's replacement as Island Professor of Chemistry. Research continued to a lesser extent in Trinidad and British Guiana.

One of the key figures to emerge in Morris's breeding program was the young Frank Stockdale, a recent Cambridge B.A. with a first in the natural sciences tripos. Stockdale had the added qualification of coming from a family of East Anglian farmers. Therefore, Stockdale was not only a talented young scientist, but he had plenty of practical agricultural experience. Morris appointed him as "Mycologist and Lecturer in Agricultural Science" in 1905, but soon he found himself working with the Barbadian breeders.[10]

In 1906, Stockdale and Morris published a "state of the art" report on sugar cane breeding in Barbados. Breeding work had begun in the 1880s, twenty years before European scientists rediscovered Mendel's research on variation and heredity. By the time of Stockdale and Morris's experiments, sugar cane research still depended upon empirical principles of selection. Stockdale and Morris selected canes by looking for variations in vegetative vigor and sucrose content. Then they took their selections and crossed them in the hope that some of the progeny might prove to be useful. Sugar cane was a relatively easy plant to cross in this way, and early, non-specialist breeders like Morris and Stockdale could produce thousands of new varieties every year. The problem was that the sheer numbers of new varieties made it difficult to select the best canes.[11]

These early breeding experiments could not predict results with any great accuracy, but Barbados breeders did develop new techniques to enhance their selections by controlling the parents. Once the cane reaches maturity, the top of its stem produces an "arrow" bearing tassels of hundreds of tiny flowers. Each flower has male and female organs, but not every cultivated cane variety is completely fertile. In natural conditions, the wind causes pollination, so that it is difficult to determine the parentage of seedlings.[12] The earliest breeding experiments relied on wind pollination, but this proved haphazard, even when steps were taken to identify the seed-bearing parent. Between 1904 and 1919, Barbados breeders developed a method called "emasculation" that had first been proposed by Bonâme in 1899. In an effort to secure the identity of both parents, a worker removed the anthers from the flowers by hand, rendering them completely female. This "emasculation" procedure was not terribly practical because some cane arrows contain more than a hundred thousand tiny flowers, and these may open over the course of ten or twelve days. The workmen had to perform the procedure with a microscope, while standing for hours on a two-meter-high platform in the tropical sun. Morris and Stockdale recognized the impracticality of these methods, and wrote that they were looking forward to a day when Mendel's laws of heredity, just recently rediscovered, might allow sugar cane breeders to develop ways to conduct their research in a more systematic and predictable fashion.[13]

At the turn of the century, the systematization of breeding led scientists to represent their canes differently. Previously, canes had evocative names, such as Bourbon or Tanna. Now, Morris, Stockdale, and breeders around the world were promoting canes with numerical names like B.147. The name of the new cane represented a significant change in thinking. On the one hand, it derived from a new, practical method of classifying canes. Faced with thousands of crosses during the early years of empirical breeding, breeders around the world went over to a numerical system of classifying their new plants. Numerical record-keeping served a practical purpose, namely, to help breeders identify canes quickly and unambiguously. But the numbers also suggested that the production of new canes was "scientific," when in fact breeders were far away from being able to predict the outcome of their crosses. The letters in front of the numbers were also significant representations. The "B" signified that cane breeders created the plant in their research facilities in Barbados, a practice that other stations followed; eventually, breeders in Mauritius would place an "M" in front of their cane numbers. Why change from the old system of naming canes after collectors and Pacific islands, which emphasized the discovery and subjection of new landscapes? The new names showed that a breeding station had a claim on a plant's future success. As research shifted from the landscape of the Pacific to greenhouses and nurseries in sugar-producing regions, it was no longer necessary to give canes evocative geographical titles; an M or a B would suffice to locate the plant spatially.

It was important to represent the new canes to the public in a favorable way. Of course, the canes had to be more practical than the old canes, with higher yields and disease resistance. But they had to be demonstrated as practical. Morris worked hard to establish the credibility of sugar cane research and the Imperial Department among the West Indian elites. Almost every year, he convened a conference in a different colony, which was attended by scientists, planters, teachers, religious leaders, and state officials. In his address to the first conference, he emphasized sugar cane breeding above other aspects of research. At one point in his speech he even brandished the new seedling cane, B.147, to illustrate its superiority to the old Bourbon cane.[14] At all subsequent conferences, sugar cane breeders had opportunities to publicize their results alongside the research of pathologists and agronomists. All of the conference papers were published in the department's *West Indian Bulletin*, which circulated throughout the West Indies and other cane-growing regions, including Mauritius. Morris recognized that planter elites in Barbados and elsewhere in the British West Indies yearned to accept metropolitan advice, but he still needed to cultivate a relationship between planters and the department in order to produce believable and useful scientific results.

It helped that Morris shared the elite's imperialist political agenda. Morris was an avowed Social Darwinist who mistrusted "the blacks." He even asked junior officers to restrict their socializing to the elite.[15] It probably helped that Morris shared the outlook of many planters and colonial administrators, whose power was closely intertwined with the department. But the point here is not to show that the science of the Imperial Department was somehow contaminated by imperialist ideology. It is to show that in the ideas and activities of Morris, the boundaries between what was "scientific," "political," and "social" were somewhat ambiguous, if they existed at all.

Despite his political views, Morris did not hold poor, "black" planters at arm's length and restrict his dealings to the elite. Morris also embarked on a strong, somewhat paternalistic effort to raise the cultivation standards of peasants on the economic and geographic margins of the British West Indies.[16] He hoped to integrate peasants more closely into the export economy, and he promoted alternative crops like bananas and citrus as a way of shoring up social stability. The *West Indian Bulletin* was filled with reports from department officials, itinerant agricultural instructors, and missionary teachers on the best ways to accomplish these goals. Many of Morris's activities presaged later efforts at what would be called "development." In fact, Morris's protégés, Stockdale and Tempany, not only went on to direct the Mauritian department of agriculture, but during the 1930s and 1940s they became the senior advisors to the Colonial Office who designed the first colonial development programs.

By the time Morris retired in 1911, he had succeeded in strengthening agricultural research institutions in Barbados and the other British West Indian colonies. Morris's knack for self-promotion and networking almost certainly spread these ideas of agricultural extension to other tropical colonies, too, but this remains a subject for further research.[17] Eventually, the Barbadian legislature created its own local department of agriculture in 1919, something that only the larger colonies had at the time. Other smaller colonies followed the example of Barbados, and the Imperial Department faded from importance. Still, the British West Indies Central Sugar Cane Breeding Station remained in Barbados. It conducted research using methods that were fundamentally similar to those employed in Mauritius and elsewhere in the cane-growing world.

But Barbados and Mauritius are different in key respects. What makes Barbados less interesting than Mauritius is that in Barbados the plantation sector dominated the economy thoroughly, making the establishment of colonial science among small planters less noticeably important. Barbados became an important site for the production of knowledge about the sugar cane plant. However, the island's social and economic structure caused more competition between labor and management on the estates, rather than competition between a large-scale and a small-scale sector as in Mauritius. Knowledge of how to produce sugar cane science and technology became a significant component of the Indo-Mauritian small planters' struggle for liberation, while it remained a smaller component of class strife among Barbadians.

The greater urgency of small planter problems may have also affected the ways in which Morris's ideas for helping peasant cultivators were translated into the Mauritian context. Morris saw it as his paternalistic mission to help peasants, and Morris's protégés, Frank Stockdale and Harold Tempany, pursued these broad objectives over the course of their careers. But when these two men served as directors of agriculture in Mauritius, they changed very little in the lives Indo-Mauritian small planters.

Creating a Mauritian Department of Agriculture

In the West Indies the Imperial Department of Agriculture flourished, while in Mauritius the Station Agronomique floundered. Once again, Mauritians took to debating the future of research institutions. Nothing concrete came of the discussions until 1909, when the Royal Commission of Enquiry visited Mauritius to investigate the island's finances. The Chamber and the Council of Government were asking Britain for a loan of £700,000 to help with agriculture and reforestation, and the wary Secretary of State for the Colonies appointed the commission to research all aspects of the economy. Over a period of ten weeks, the commission heard testimony from ninety-nine witnesses.

The commissioners recognized the importance of agricultural research and made a point of asking about it. Philippe Bonâme testified that if a Department of Agriculture with a staff of specialists were created, it would complement his own research. The most prominent members of the Chamber also testified in favor of creating a Department of Agriculture. The 1909 commission agreed with the Chamber, but like the Colonial Office, it thought any new government department should serve everyone in the colony, not just the owners of the large sugar estates. As one commissioner put it while questioning Bonâme, the new department would "make experiments with all kinds of new products, and assist everybody who was an agriculturist in the island."[18]

The Chamber openly supported the notion of such a public, accessible Department of Agriculture, but during lengthy discussions with the governor, the members worked behind the scenes to maintain their control over agricultural research. In 1911, the Chamber voted a resolution outlining its vision for the new research institution. The members were impressed with the accomplishments of the Imperial Department of Agriculture of the West Indies, but they disliked how the Imperial Department obtained its budget: it was financed through general tax revenues and therefore it was firmly ensconced within the budgetary control of the imperial government. The Chamber claimed disingenuously that such an institution would be too big and complicated for Mauritius, even though the island routinely produced as much sugar as all the islands in the British West Indies.

The Chamber promoted a small department that would not be a part of the civil service. They wanted to finance it through a tax on exported sugar, much like the Station Agronomique had been financed, because if the sugar industry paid for the new department, then it would have a greater say in its affairs. The Chamber argued that the department should have a pathologist, an entomologist, an agricultural chemist, and a botanist, one of whom would serve as director. A board of Chamber members would advise on the department's research agenda, much as they had done for the Station Agronomique.[19]

In other words, the Chamber was proposing to dominate a government department. This was brazen but it was also the next logical step in Mauritian agricultural research. Still, the Chamber's proposal evoked disbelief in London. The governor of Mauritius, Cavendish Boyle, was about to retire and may not have wished a last-minute confrontation with the local planters. He showed his overall support for the Chamber in his correspondence with the Colonial Office. Upon reading Boyle's letter of 23 February 1911, in which the governor described the negotiations concerning Bonâme's possible future employment in the new Department of Agriculture, the Secretary of State's assistants assailed the governor and the Station Agronomique Committee in an attached minute. One remarked, "It is no use lancing this Boyle when it is just about to burst," and another responded, "I hope we shall be as successful in getting rid of Mr. Bonâme."[20]

After Boyle's retirement, the Chamber continued to negotiate with the interim government and the Colonial Office about the new department's institutional arrangements. The Colonial Office took a dim view of the Station Agronomique's performance. Many members of the Chamber, including Leclézio, hoped that the station's staff could be incorporated within the new department, with Bonâme occupying the position of assistant director. Such a department would retain some independence from the government, even though it would receive its money from the colonial treasury. However, the Colonial Office aimed to create an independent state department of agriculture, rather than an organization under the Chamber's domination. In response to a letter from the "officer administering the government" in the governor's absence, a minute circulated within the Colonial Office that sketched an outline of how London expected the Mauritian government to handle the Chamber:

> The plan of the Station Agronomique is only more absurd than that of the Chamber of A. The Director if he were enterprising would be crushed by the weight of his committee, & the dept. would take Govt money & be run in the interests of the planters, & all representatives from the Govt would leave the committee unmoved. Moreover Mr Bonâme would always be supported as against the new Director.
>
> I think that the only thing to be done is to press for the recommend[ns] of the Royal Comm[n] & nothing but them; if we oppose a blunt negative to every wild goose scheme put forward by the planters, Mr Bonâme (being now 59 or 60) will in due course be dropped, & in any case the withdrawal of the govt subsidy (derived from a tax on exports of sugar) can always be threatened.[21]

The Colonial Office policy to prevent the Chamber from interfering in the new Department of Agriculture immediately became the subject of negotiation in Mauritius. At first the new governor, J.R. Chancellor, resisted the Chamber's plans to dominate the proposed Department of Agriculture.

However, during a meeting with a delegation from the Chamber he approved the substance of their plan.[22] Bonâme would be named second-in-command, and a "purely advisory" committee of nine members of the Chamber would supervise the department's work. The staff of the station would be retained as employees of the department, including Bonâme's assistant, Pierre de Sornay. Other Franco-Mauritians would be appointed to the department's staff, including Donald d'Emmerez de Charmoy, the entomologist in charge of the Mauritius Institute, and Henri Robert, the Chamber's statistician.

Once again, the Colonial Office was appalled. One senior bureaucrat argued that a department under the control of the Franco-Mauritian elite would produce politicized, shoddy science. He suggested that it would have been better simply to wait a few years until Bonâme "disappeared," preferably without a government pension. He added hopefully that "even Mr. Leclézio cannot last forever." Mainly he expressed concern that the proposed arrangements "would leave the Director of Agriculture largely at the mercy of Mr Leclézio," and outnumbered by "Mr Bonâme, the deposed, and Messrs de Sornay and Pougnet - all three of them almost certainly Mr Leclézio's men." He assailed the proposed Franco-Mauritian staff members viciously in an effort to undermine their credibility:

> What position would be taken up by Mr D'Emmerez de Charmoy, the ex-Curator of a slovenly and useless Museum . . . I cannot say. As for Mr Robert . . . he is a local journalist and was at one time Editor of the Journal de Maurice in which capacity he duly supported his leaders Messrs Leclézio and Newton. He then became Secretary of the Chamber of Agriculture, and when in 1909 the Chamber started what they call their Bureau de Statistiques he was put in charge of it. I annex one of his productions . . . it is exactly the sort of thing which we don't want the Department of Agriculture to produce—long, tedious, and useless. He is I believe intelligent - a Mauritian journalist has to be if he is to escape the Poor Law Department and the Prison Department.[23]

The Colonial Office enlisted the support of Kew's new director, Col. David Prain, in pressing for an independent department of agriculture. The Colonial Office did not wish to undermine the new governor's credibility with the sugar industry, and thought it might be easier for him to reverse himself and deny the Chamber's requests based upon "technical advice."[24] Prain wrote to the Colonial Office that the Chamber's proposed advisory committee was so "detrimental to the efficiency of the projected Department of Agriculture," that it must have been "calculated to impair

[its] efficiency." In addition, he pointed out that if the new department was to work to improve *all* agriculture on the island, then it should draw its funds from the treasury and not from a tax imposed on one industry. The Colonial Office forwarded Kew's arguments to Chancellor, adding that under the terms of the governor's proposal, it would be better not to have a department of agriculture at all.[25]

Given the choice between contradicting London or the Mauritian sugar industry, Chancellor erred on the side of local expediency while making some concessions to the Colonial Office. This was only natural: British governors had been ruling Mauritius in this way for the past hundred years. In a letter to the Secretary of State, Chancellor argued strenuously against postponing the creation of the department, stating that it would be "detrimental to the general interests of the Colony."[26] He did grant that no appointments should be made to the department until the new director could assess the competence of the applicants, although Bonâme would definitely become the assistant director.[27] Finally, in his most important point, Chancellor affirmed that the advisory board would not control the department. He conceded that the Governor, and not the Chamber, should appoint the board in order to "remove the danger of excessive representation of particular interests."[28] Chancellor believed firmly that having such a board would be necessary in order to gain the cooperation of the sugar industry. The credibility of scientific institutions depended heavily on community oversight, although Chancellor did not see the community extending beyond the Mauritian elite. He wrote that the "confidence and willing cooperation of agriculturists is essential to success. To attempt to push through a scheme in the teeth of opposition would be to court failure from the beginning and might lead to the continuance of the Station Agronomique independently of the new Department."[29]

In the end, Governor Chancellor brokered a compromise that balanced local and metropolitan interests in the time-honored tradition of colonial Mauritian politics. The Colonial Office backed down because Chancellor had reassured them that the department would be independent of the Chamber. The Colonial Office also did not wish to undermine a new governor's decisions in the eyes of the colony.[30] In the final agreement, Chancellor accepted the notion of the treasury supporting the department, rather than a special tax on the sugar industry. He also revised the advisory committee by giving the governor the power to appoint its members. The Colonial Office approved these changes.

In 1913, the Department of Agriculture finally came into being, absorbing the Station Agronomique, the gardens, and the Chamber's statistical bureau.[31] The Chamber had succeeded in obtaining a department of agriculture, but its members were disappointed when they learned that they would not have as much institutional and fiscal leverage as they had sought.[32] The advisory committee was formally toothless, but the Chamber remained an effective lobbying organization in its own right. The sugar industry no longer supported research directly through a special tax, but the estates still dominated the economy and contributed heavily to the colonial treasury. The franchise was limited to the elite, and many colonial legislators belonged to the Chamber. For these reasons, the Chamber was able to influence the Department of Agriculture's production and distribution of science and technology.

The Department of Agriculture under Frank Stockdale, 1913–1916

In 1913, cooperation between the sugar industry and the colonial government allowed the new Department of Agriculture to get off to a running start. The Secretary of State for the Colonies appointed Morris's thirty-year-old protégé, Frank Stockdale, to be its founding director. Stockdale, who had risen to become the assistant director of agriculture in British Guiana, was an experienced cane breeder, among other things, and was one of the rising stars of British colonial science.[33] He built the new, interdisciplinary department in Mauritius while implementing many new ideas, including a veterinary service, as well as extension services and cooperative credit societies for small planters. Stockdale had a broad vision for colonial science, in which science, society, and economy were inseparable. Many years later, an official of the Colonial Office named Charles Jeffries would write that Stockdale personified "the application of scientific knowledge to the economic and social development of the colonial territories." Jeffries called attention to Stockdale's ability to "view agricultural problems in relation to the general economy and social conditions of the territory concerned." Stockdale was not only an accomplished scientist, but he was also a practical man with plenty of agricultural experience from his childhood in East Anglia. He also had the kind of social qualifications that may have stood him in good stead with the Mauritian elite: during his tour of duty in the West Indies he married Annie Packer, the daughter of a Barbadian sugar planter. Stockdale was also known to be an affable, agreeable fellow.[34]

For all these reasons, it is not surprising that Stockdale navigated the murky waters of Mauritian politics with great finesse, unlike Bonâme. He

reported to Kew that when he arrived in Mauritius, he found "vested interests galore," and the Chamber seemed "quite friendly disposed," inviting
him to speak at one of their meetings.[35] The Chamber greeted him in a
spirit of cooperation and change. Possibly for the first time in the
organization's history, several Indian estate owners and planters attended
the meeting, indicating that the members wanted to show themselves in a
more broad-minded light than usual.[36]

Stockdale immersed himself in local agriculture. He spent his first
four months as Director of Agriculture touring the countryside. In his personal letters he complained about being "stuck" in this "far from desirable"
place, but on the surface his interest in island affairs probably impressed
local farmers, even though he could not speak French or Kreol. The farmers were accustomed to British administrators taking a hands-off approach,
and Stockdale represented a welcome change of pace. Stockdale seems to
have recognized that he could increase the production and distribution of
agricultural research through cooperation with the estates. For their part,
the estates recognized that they could achieve higher productivity through
cooperation with Stockdale. The Chamber and the Department of Agriculture had many reasons to enlist each other's support.[37]

Central to this new relationship was the expansion of agricultural
research out of the laboratory and into the fields. Stockdale used some of
Bonâme's ideas in his plans for the future activities of the Department of
Agriculture, but he deprecated his predecessor as being from "the old test-
tube school."[38] Stockdale outlined his plans in a report to the Board of
Agriculture. The sugar industry needed to obtain new canes for the sake of
security and also because some estates lacked suitable varieties. The department would continue to import canes while conducting basic breeding
experiments. The chemist would also have to analyze soil and climatic conditions in different districts, in order to understand which manures and
cane varieties worked best. The department would use Réduit as a central
experiment station, while testing new varieties and agricultural practices at
subsidiary stations around the island, because of the island's varying micro-
climates and soils.

Under Stockdale's plan, the new department would also advise sugar
growers on agronomy, emphasizing how to use implements better.
Implemental tillage would help the sugar estates greatly. As late as the decade of the 1900s, rocky soil and cheap labor meant that most growers did
not even use plows. When indentured immigration ended in 1910, they
began to reconsider their usual labor-intensive practices. By 1914, 12.5
percent of the land under cane was being plowed, even as the estates were

calling for the resumption of the "coolie-trade."[39] Another important part of the department's agronomy program would be to convince growers to maintain special plots of selected canes to use as cuttings for planting, rather than relying on random cane tops that were the remnants of the harvest. This practice would extend the economic life of any given cane variety because poor canes would be selected out. The department would also conduct research in entomology and plant pathology with an eye toward controlling the *Clemora smithii* beetle as well as fungoid diseases.[40]

Just as the state was beginning to promote a broadly based agricultural policy as outlined in the 1909 commission's report, the Chamber intruded upon Stockdale's appointments to the new department. A Board of Agriculture comprising members of the Chamber and government officials oversaw the department's work. Stockdale wrote that he "swallowed" the board's proposals for staff. For example, the Chamber was willing to subsidize a department statistician, but insisted on transferring its own statistician, Henri Robert, to the post. Stockdale wrote that the statistician "will have to dance to my tune," but given the historic role of private industry in Mauritian public institutions, this was not entirely likely and showed Stockdale's inexperience. The Chamber, the governor, and the secretary of state insisted on retaining Bonâme as chemist and assistant director, but the old Frenchman spurned their offer and decided to retire rather than work under the young Englishman.[41]

The early debates over the staffing of the department call attention to an important issue in the history of Mauritian agricultural science, an issue that raises interesting questions for the history of colonial science. Beginning in 1913, British expatriates formed the majority of the agriculture department's senior professional staff. But Stockdale also appointed Mauritians to senior and junior positions: Donald d'Emmerez de Charmoy and Pierre de Sornay, whom the Board of Agriculture asked Stockdale to appoint, and Charlie O'Connor, who was already working for the Forest Department. Over the course of their careers, these three Mauritians influenced the department's trajectory significantly, showing how the local scientific tradition could merge with a colonial agriculture department.

Donald d'Emmerez de Charmoy became the department's entomologist, and his appointment worked out well. He was very much a local scientist: in fact, he had no formal scientific training, and had learned whatever he knew by reading and collecting. He began his career as the assistant to Albert Daruty de Grandpré, the curator of the Mauritius Institute, and by 1913 he had risen to direct the Institute himself. At the same time, he also gained important experience working with the British, directing a government anti-malarial project from 1907 to 1913. From the time of his appointment in the department in 1913 until his death in 1930, he conducted basic research and published extensively on both entomology and animal husbandry. During the 1920s he also served occasionally as acting director of the department, and he was the director during 1929–1930.[42]

Charlie O'Connor was also an autodidact from an elite family. His father had been the government's Inspector of Immigrants, while his mother belonged to the Franco-Mauritian branch of the Dupont family. He loved horticulture, and after graduating from secondary school at the Royal College of Curepipe he went to work for the Forest Department, first as the overseer of the Curepipe Botanical Gardens, and later as the superintendant of the governor's gardens at Le Réduit. Stockdale appointed him as the chief overseer of all the gardens in Mauritius as well as caretaker of the department's experimental grounds. After service in the First World War, he studied horticulture at Kew, spent several years as a government horticulturist in Zanzibar, then returned to Mauritius in 1923 to take up the position of Senior Agricultural Officer. He held this position that until his retirement in 1944, and like d'Emmerez he served as acting director of agriculture, once in 1937–1938.[43]

By contrast, Pierre de Sornay had better formal qualifications than d'Emmerez or O'Connor, but he had difficulty working under the British. He too was a product of the local scientific tradition, but he was also a Francophile. After graduating from secondary school in Mauritius in 1893, he studied as a seminarian in Avignon for four years. He lost his vocation and returned to Mauritius in 1897, where he studied agricultural chemistry under Bonâme. He became assistant director of the Station Agronomique, and seems to have become Bonâme's protégé. His French training and his alliance with Bonâme seem to have contributed in some way to his dissatisfaction with his position as chemist to the department of agriculture, and he resigned in 1916. He remained active in local science, serving as a consulting chemist to the sugar industry in Mauritius as well as to the French colonial government in Madagascar. He was a prolific writer as well as a member of many of the local learned societies, and over the course of his career he spoke out against the department of agriculture on a number of issues.[44]

Most of the Mauritians on the staff were either Franco-Mauritians or assimilated Creoles, the ethnic groups that dominated the sugar industry. This was the beginning of a trend in the department of agriculture. The early appointments of Mauritians showed that the Chamber preserved its influence over state institutions and agricultural science. But this was not just a case of political influence over scientific research. Mauritian agriculturists regarded themselves, often correctly, as better qualified to answer questions in local agronomy than outsiders. The island had a long history of scientific research dating back to the eighteenth century, which was now being perpetuated through affiliation with state institutions. The department could improve results and gain credibility by hiring local scientists. In Mauritius, colonial science incorporated both local and metropolitan influences.

To some extent, Stockdale depended upon locally trained people, but he also put into place a program for training Mauritians more formally in the agricultural sciences. Previously, Bonâme had accepted three students each year to learn sugar chemistry at the Station Agronomique. When he first arrived in Mauritius, Stockdale reviewed the state of agricultural training. In 1914, he founded a School of Agriculture that was placed under the aegis of the Department of Agriculture. Stockdale instituted a three-year syllabus in agriculture, which included courses in botany, chemistry, entomology, and meteorology. Stockdale hoped that the School would train the future staff members of the sugar estates, but that it would also provide a cadre of local scientists for the department. The School was housed in the department's Réduit headquarters, and consequently it had room only for six students.[45] It was a small start for formal training in Mauritius, but it was a significant step toward the merging of local and metropolitan science.

The Department of Agriculture under Harold Tempany, 1916–1929

During Stockdale's tenure as Director of Agriculture (1913–1916), the department and the Chamber established a close working relationship. The Chamber was pleased to see the colonial government conducting agricultural research and many people expected good results. The Chamber lobbied the department mainly to obtain new cane varieties, an issue that received plenty of attention. In 1914, the Chamber felt that the Tanna canes were starting to degenerate and asked the department to study the problem.[46] Stockdale did so, and decided that the department needed to import more canes. The Chamber proposed acquiring some of the new Javan canes, but since the Dutch bred these to be used only as virgins, Stockdale recommended obtaining samples of Queensland's newly collected canes from New Guinea and New Caledonia, which had greater ratooning power. In any event, from 1914 to 1916 the department acquired canes from around the world. Stockdale also proposed placing new canes in quarantine to avoid introducing pests; since the 1890s, the Chamber had been promoting this new procedure for the island.[47] At the Chamber's urging, d'Emmerez also began work on *Clemora smithii*. He experimented with insect parasites of beetles imported from Africa and the West Indies, and the Chamber paid for him to take an insect-collecting trip to Madagascar.[48]

In 1916, Stockdale left Mauritius. It is not known if this was a case of "push" or "pull." On the one hand, the Colonial Office may have had bigger plans for him: his next posting was Ceylon. But in his correspondence he complained about "all the racial bickerings" in Mauritius.[49] In any case, Stockdale's successors followed his policy of cooperation with the Chamber. In the absence of detailed metropolitan direction, this was the easiest path to take. The Department of Agriculture responded to local pressure because this is what the British colonial government had been doing in Mauritius for a century. In addition, the laissez-faire Colonial Office did not yet have a clear, consistent agricultural policy of its own. It merely encouraged the introduction and production of some new crops and expected local governments to raise revenues, which came from agriculture in most colonies.[50]

The Colonial Office appointed Harold Tempany as the new Director of Agriculture, a choice that indicated a desire for continuity. This is certainly what the Chamber wanted, as they made clear in their lobbying efforts during the transition in leadership.[51] Like Stockdale, Tempany was educated in England and began his career under Morris in the Imperial Department of Agriculture in the West Indies. But in some ways Tempany was more formidable than Stockdale. He was educated at University College, London, where he received his doctorate in chemistry, and he was a fellow of the Royal Institute of Chemistry. He began his work in the West Indies in 1903, and over the course of thirteen years he had advanced from the rank of Assistant Agricultural Chemist in the Leeward Islands to Superintendant of Agriculture in the Leeward Islands. Therefore, Tempany had already had the experience of running another department, and he was also three years older. Tempany was just as dynamic and strong-willed as Stockdale, but while Stockdale had a genial manner Tempany could be politely brusque.[52]

Under Tempany's stewardship, the department became even more closely associated with the interests of the Chamber. At first, Tempany continued Stockdale's work while learning about the Mauritian situation. His earliest successful joint venture with the sugar estates involved increasing the use of plows. Since 1914, the department had been encouraging implemental tillage on the estates in order to decrease their dependence on a shrinking pool of laborers.[53] However, wartime shipping shortages limited the supply of plows and other tools. Tempany recognized that the war raised sugar prices artificially, causing Indo-Mauritian small proprietors to expand their land under cane. This in turn depleted the estates' labor reserve.[54] In 1917, he predicted that when peace returned, prices would fall sharply. He argued that the department and industry's "attention should be directed to effecting economies in production whenever possible, both in the factory and the field."[55]

When the war ended, Tempany and the Chamber were faced with a rapidly changing economic situation. The Mauritian sugar industry continued to depend on the good graces of the British government for a stable market, because Whitehall's postwar policies continued to give colonial sugar producers preferential access to the home market. Starting in 1919, the British government created "Imperial Preference," a system of tariffs favorable to colonial cane sugar producers. The Labour government of 1924 reduced the preference as a way of relieving British consumers, and Mauritian sugar reverted temporarily to its old Indian market. However, the Conservative government of 1925 restored the full preference, which remained in place up until the Second World War.[56]

Mauritius remained vulnerable to changes in the marketing system. In 1919, the majority of millers formed a central marketing organization called the Sugar Syndicate in order to increase their leverage in negotiating a price. But even this was not enough protection. In 1928, British sugar refiners convinced their government to raise the duty on imported pure-white crystal sugar. This encouraged the colonies to produce raw sugar to be refined in Britain, rather than to refine it themselves. This change hit Mauritius especially hard, because factory owners had invested considerable sums in improving their equipment to produce pure white sugar.[57]

During the 1920s, the British government also began to show a greater interest in investing in colonial plant science, a trend that continued through the 1940s. In 1919, the Colonial Office established a Colonial Research Committee that distributed funds to scientists in the colonies. Between 1926 and 1933, the Empire Marketing Board provided funds for research to many government scientists. The Colonial Development Act of 1929 provided additional funds to state research institutions, to be succeeded by more substantial money after the passage of the Colonial Development and Welfare Acts after 1940. Staffs of colonial departments of agriculture expanded, in some cases doubling and tripling. Many of the new officers received their training in new imperial institutions, such as the Imperial

College of Tropical Agriculture in Trinidad, as well as at new university programs in England geared to colonial agronomy.[58] In 1929, the Colonial Office created the position of agricultural advisor to the secretary of state, a post first occupied by Stockdale, and in 1935 it unified all its agriculturally oriented civil servants into a new Colonial Agricultural Service.

One retired member of the service, Geoffrey Masefield, has written that even as London took a greater interest in colonial agricultural science, colonial agricultural officers identified strongly with local issues and problems. All agriculture departments promoted crops and established research facilities, but a specific agricultural policy for all colonies never existed. Most agricultural officers subscribed to *laissez-faire* economics while encouraging scientists to educate farmers. In remaining sensitive to local issues, agricultural officers were following broader trends in colonial policy. They derived moral justification for their work from Lord Lugard's *Dual Mandate in British Tropical Africa*, required reading for all trainees; Lugard argued that both "natives" and Europeans should benefit from colonial rule. In most colonies agricultural officers supported various systems of "indirect rule," consulting with local leaders before implementing policies. In later years, most also took an interest in diversification, soil conservation, and nutrition. In practice, agricultural officers had to tailor these general policy outlines to local social, economic, and ecological conditions. Local elites attempted to gain productive advantages by influencing the production of agricultural science and technology. Colonial taxation generated most of the revenues for departments of agriculture, giving colonial populations a further stake in research. Moreover, expatriate British agricultural officers were always thin on the ground in the colonies. They depended upon local cooperation for most of their initiatives.[59]

Under Tempany, Mauritian staff members played a significant role in the department. The names of staff members listed in the department's annual reports tell an interesting story: about half had French names and half had British names, while the only Indian names listed held clerical positions in the cooperative societies or were overseers in some of the gardens and experimental plots. Tempany continued to employ Donald d'Emmerez de Charmoy and Charlie O'Connor, two autodidacts who represented the best qualities of the local scientific tradition. And while most of the department's senior officers came from Great Britain, Tempany also added several Mauritians to his staff, some of whom went on to have long and distinguished careers in the department. Each, in his own way, represented the changes in local scientific training.

Gabriel Orian, who became Tempany's plant inspector in 1925, was a good example of the older, autodidactic Mauritian agricultural science. He was a Franco-Mauritian who had graduated from secondary school at the Royal College of Curepipe in 1915. He went to work at the Bois Rouge sugar estate, where he learned about science on the job. Even though he lacked formal training, he developed a strong interest in plant pathology, which is why Tempany hired him. Between 1925 and his retirement in 1955, Orian published thirty-three articles on sugar cane diseases, and rose from plant inspector to become the department's chief pathologist.[60]

If Orian represented the older tradition, then André Moutia represented a newer one. Moutia was a Creole who had also graduated from the Royal College, but he went on for post-secondary training at the new School of Agriculture. He received an honors diploma in 1922, then went to work immediately for the department. He served as the assistant entomologist under d'Emmerez, and over the course of thirty-eight years of government service he published thirty-one articles, he received further training overseas, and he traveled extensively to collect insects. In fact, it is curious to note that even with his many accomplishments, he was not appointed chief entomologist until 1957, three years before he retired.[61]

But Tempany also took steps to change the kind of agricultural training that Mauritians received. He recognized that it would be important for Mauritians to obtain formal agricultural training of the sort that Bonâme had begun to give at the Station Agronomique. To this end, Tempany founded a College of Agriculture in Mauritius, which revealed his interest in research and education as well as his shrewdness as an administrator. In 1920, he predicted that the postwar sugar boom would not last long. While the sugar industry could afford it, he persuaded the Chamber to allow the government to raise a windfall export tax on sugar, Rs.40 per metric ton. The Chamber administered the proceeds, which became the Development and Improvement Fund for Mauritian agriculture. This tax resembled the Station Agronomique tax in that it allowed the Chamber to receive funds from all sectors of the sugar economy for projects that benefited the estates most directly. Tempany convinced the Chamber to spend this money to create a government College of Agriculture at Réduit, which would train Mauritians in agronomy and sugar factory engineering. The Legislative Council approved a new export tax of three cents per 100 kilograms of sugar for the school's maintenance. The college opened in 1922 under the aegis of the department, admitting approximately five students each year to its three-year course.[62] The department benefited from the college, not only because it provided trained graduates; Tempany used college lectureships as a crafty way of hiring extra department staff, without having to ask the governor for more departmental funding.[63] But he also made it possible for more Mauritians to receive formal, European-style agricultural training.

One of the earliest college students, Jean Vinson, was a Franco-Mauritian who went on to have a great influence on science in Mauritius. Under the tutelage of d'Emmerez, Vinson came to love entomology. He graduated in 1926, and after two years of working for a sugar estate, he joined the department as an entomologist. He worked in that capacity with André Moutia until 1941, when he joined the Mauritius Institute as a curator. Eventually he rose to direct the Institute from 1954 to 1966. Vinson published thirteen articles on entomology, and with Moutia he increased the general understanding of insects in Mauritius.[64]

The careers of Moutia and Vinson show some of the ways that the new agricultural education was having an effect on sugar cultivation. The department drew some of its most talented recruits from the college, and the sugar estates hired many graduates, too. But there was also a notable absence at the college. Jagadish Manrakhan has prepared a list of all the college's graduates, from its inception as the School of Agriculture in 1914. From 1914 to 1942, there is not a single Indian name on the list; from 1942 to 1952 there are three Indian names; and after 1953 there are considerably more. The list indicates that the early college catered exclusively to Franco-Mauritians and Creoles.[65] This would be consistent with the department's tendency to ignore the increasingly influential Indo-Mauritian small planters.

Changes in education played a significant if restricted role in the department's broader efforts to improve Mauritian agriculture. Tempany also sought other ways to collaborate with the sugar industry. During the 1920s, the department cooperated with the Chamber to deploy new technologies to reduce the estates' wage bill. By 1919, with the influenza epidemic removing even more laborers from the workforce, Tempany was hoping that the estates could solve their labor shortage by increasing implemental tillage, rather than by making estate work more attractive to laborers. He reported that "discussions" had taken place, presumably with the Chamber, on the subject of introducing internal combustion tractors.[66] In 1922, he arranged for an importing firm to demonstrate different types of tractors at an estate near Réduit. Fifty-eight estate managers and staff members attended, along with the acting governor, and following the show tractor purchases increased dramatically: by the end of 1922 there were forty-seven operating on the island. After further department demonstrations in 1923, the number rose to 135. Tempany sided openly with the estates, believing that mechanization had gone "a long way toward bringing round the local labour to a saner view of the altered conditions of the sugar industry." Rocky soil had discouraged many estates from using tractors, but thanks to the mid-1920s contraction of the small planter sector estate labor became plentiful and mechanization was no longer necessary. In 1926, the number of tractors in use fell to 117, and by the 1930s, with sugar prices at their nadir, there were hardly any in use.[67]

Tractor trials and the founding of the College of Agriculture helped the estates a great deal, but the department's chemistry division quietly accomplished much more for sugar estates. Understanding soil composition is one of the least glamorous but most essential aspects of agricultural science, and for this reason the department's chemists began to conduct a soil survey of the island in 1914. As far as the estates were concerned, the chemistry division's most important work was to analyze soil samples from their fields, as well as to conduct experiments on the effects of different fertilizers. Some estates allowed the department to establish experimental plots in their fields and benefited directly from government research. Other estates paid the department's laboratory to conduct analyses at the rate of one or two rupees per procedure. This was a nominal fee for a large proprietor, considering the potential benefits, but it would have been expensive for a small planter, none of whom had access to this program. Soil analysis proved very popular among the estates and it boomed under Tempany: in 1913 the department performed 519 procedures, and 309 in 1914, but in 1924 the figure rose to 4997, and in 1929 it tapered off at 1439.[68] The department also conducted extensive fertilizer trials, using manure, guano, molasses, and various chemical fertilizers. During the 1910s and 1920s, these took place at the departments Réduit headquarters and on the grounds of eleven estates. By contrast, it was not until 1937 that the department analyzed small planters' soils.

Yet it was cane breeding, rather than soil chemistry, that became the most important way in which the department and the estates collaborated. As always, the production of new cane varieties held great rewards, while a lapse in their production could jeopardize the entire sugar industry. When Tempany first arrived, he concentrated on making cane breeding trials and importations more systematic, conducting experiments at Réduit, Pamplemousses, and six sugar estates, with the idea of adding one more government experiment station and three more estates in the near future. During the First World War, the shortage of shipping interrupted cane variety imports, impelling the department to be even more self-sufficient than usual.[69]

Tempany remarked that the Mauritian sugar industry wanted new varieties because of "somewhat widespread anxiety that, in certain localities, many of the standard varieties of cane at present cultivated are showing signs of deterioration." Like Horne before him, Tempany blamed defective cultivation techniques, but the department's efforts to improve the sugar industry's field methods were not stemming the tide.[70]

During the 1910s and 1920s, the conventional methods of the department's entomologists and pathologists failed to rid local canes of diseases and pests. There were many pest problems, including an increase in some bacterial diseases. The most persistent menace was the *Clemora smithii* beetle, also known as "Phytalus," which attacked the roots of canes and affected the weak-rooting Tanna canes especially heavily. The department preferred using biological controls against *Clemora*, because trials with insecticides proved inconclusive. Their anti-*Clemora* efforts consisted mainly of releasing wasps and toads imported from the Caribbean, Africa, Madagascar, and the Middle East. Agricultural officers visited estates to study the insect and encouraged all growers to cut the larvae out of canes in heavily infested areas, although in 1938 the department was to prove that this practice was utterly worthless.[71] The Chamber supported the department's work. In 1920 it convinced the government to levy a special tax of two cents per 100 kilograms of sugar exported to create a fund that would pay for gangs of laborers to destroy the beetles.[72] However, the insect problem only grew worse and by the end of the 1920s approximately one-fourth of the area under cane was losing about seven tons of cane per hectare, when the average estate could expect a harvest of fifty tons per hectare, and a small planter twenty-five.[73]

Tempany's efforts to select and import new cane varieties complemented the work of the department's entomologists and pathologists, but no new variety emerged to replace the Tannas. During Tempany's tenure (1917–1929), the department imported forty-six varieties from seven different cane-growing regions.[74] In addition, it intensified its own empirical selection program. The breeders could produce tens of thousands of new seedling varieties in any given year, depending on the weather. During the three years after germination, trials at Réduit and Pamplemousses winnowed the field to perhaps a dozen candidates that were then tested on approximately a dozen different estates.[75] The department's new sugar technology division, founded in 1923 to help the estates' factory operations, joined in analyzing the results. In the end, the department distributed the most successful varieties to any estate that requested them. Most of the estates now had their own experimental nurseries. In 1918, a typical year, thirty-nine estates received cuttings of thirteen varieties.[76]

Sugar cane experimentation was spreading throughout the country-side, coordinated by the Department of Agriculture with the backing of the Chamber. Much of the spread was due to the efforts of the department, but Mauritian planters also retained their own interest in cane breeding. Two planters made a reputation for selecting seedling canes: Noë Férré of Rivière des Anguilles and Eugène Fleuriot of Rich Fund. Still, during the 1920s no locally bred canes won widespread acceptance. Only one new cane attained any popularity, a Barbados hybrid called BH10(12). It is significant that Tempany introduced this cane at the suggestion of a Mauritian planter, Adrien Wiehé.[77] Gradually, BH10(12) gained some acceptance. Growers planted it during the late 1930s, and by 1940 it even accounted for 40 percent of the land under cane. (See Table 1.) However, it was hardly the cane the Chamber was seeking to renew the sugar industry: it performed well only in the drier, western and northern parts of the island, even though in Barbados it had performed well in high-rainfall areas. BH10(12) was also highly susceptible to mosaic disease, which was present in many other cane-growing countries. It was still easy to introduce diseases; in fact, in 1923 the department had almost introduced mosaic to Mauritius by accident.[78]

Toward the end of the 1920s, the lack of varieties once again spurred the Chamber to deepen its involvement in the Department of Agriculture. In April 1927, Tempany convened a conference of all the organizations involved in the sugar industry. The Chamber used it as a forum to increase pressure on the department to provide new varieties, because sugar producers felt the squeeze of declining prices. Representatives of the government and the industry agreed to establish a Sugar Industry Reserve Fund, which the governor signed into law later that year. The fund would be raised from a tax of four cents on every 100 kilograms of sugar exported and would support scientific research. The president of the Chamber would act as chairman of the fund's committee, made up of the Director of Agriculture and seven prominent sugar industrialists whom the governor would appoint. The small planters' sector would once again contribute taxes to research, but would have no voice on how to spend the money. The people attending the conference also resolved to study specific problems at later meetings, and in 1928 the Chamber appointed a committee to study cane diseases and varieties. It called on the colonial government to create a new center for raising seedling canes. It also urged the department to import six well-known foreign varieties, including Java's POJ2878.[79]

Tensions between Tempany and the Chamber were on the rise as the department's efforts in cane selection, entomology, and pathology met with

Table 1

Principal Cane Varieties Cultivated in Mauritius, 1915–1955, Expressed as a Percentage of the Total Area under Cane on Sugar Estates

	1915	1925	1930	1935	1940	1945	1950	1955
Tanna vars.	56	63	57	48	29	5		
Perromat varieties	19	16	10	7	1			
DK74	5	9						
"D" varieties.*	6	6	16	16	1			
BH10 (12)			2	15	40	27		
POJ2878			1	5	2			
M134/32					2	37	91	74
Other "M" varieties.†				1	5	7	6	5
Ebène 1/37								15

* Note: Varieties beginning with the prefix "D", such as D109, came from Demerara, British Guiana.

† Note: "M" is the prefix used before the numbers of canes bred by the Mauritian Department of Agriculture, 1913-1929, the Sugarcane Research Station, 1930-1953, and the Mauritius Sugar Industry Research Institute, 1954-Present. The number after the slash signifies the year in which the cane was bred.

Sources: Maxime Koenig, "Census of Cane Varieties Grown in Mauritius in 1933." Colony of Mauritius. Department of Agriculture. Bulletin No.7, Statistical Series. Norman Craig, "The Spread of New Varieties of Sugarcane in Mauritius." Colony of Mauritius. Department of Agriculture. Sugarcane Research Station Bulletin No.17, 1940. Guy Rouillard, *Historique de la canne à sucre à l'île Maurice* (Port Louis: MSM, 1990), pp.28-30.

little immediate, demonstrable success. In December 1927, Tempany addressed another conference on the subject of the deterioration of the Tanna canes. He stated that, "the extensive work that has been performed by the Department of Agriculture in raising and trying out new seedlings . . . has not, I think, received so far all the attention it deserves from planters."[80] Henri Leclézio, now the grand old man of the Chamber and Mauritian politics, retorted that the department should import POJ2878, the variety that had saved the Javan sugar industry. Pierre de Sornay, Bonâme's former protégé, suggested a different cane from Java, POJ2725.[81] If they thought they had a better chance of success with Javan canes than with canes from the local department, then they were probably casting a serious vote of no confidence in the department's own breeders. As one hundred years of cane imports had demonstrated, varieties selected and bred for one region rarely prove commercially successful when transplanted to another.

Tempany was sensitive to the Chamber's demands and he redoubled the department's efforts to find new varieties. As part of his approach, he put in place new institutional arrangements for agricultural science. In 1928, he opened four new subsidiary cane selection experiment stations around the island and reported that "a great extension was given to the propagation of the most promising cane varieties." He launched a public relations campaign as well, holding a "field day" to show off the department's new varieties at the Pamplemousses gardens. Like John Horne before him, Tempany arranged for the most influential planters to witness his canes and hear his representations. One hundred people interested in the sugar industry attended, and in Tempany's words, "every effort was made by officers of the department to interest the visitors in the important work of cane selection and propagation . . . They were shown round the experimental plots by the director of agriculture and his assistant. This meeting was a great success, the chief result being that the visitors impressed by the many promising varieties they saw, sent in numerous applications for planting material." The department also acceded to the Chamber's request and imported POJ2878 and the other foreign varieties.[82] Under pressure, the department's scientists made extra efforts to reach out beyond their headquarters to the estates, and to the estates alone.

Most importantly, in 1929 the private sector's desire for better canes persuaded the state to cobble together the funds to create a new branch of the Department of Agriculture, the Sugarcane Research Station (SRS), which was to last from 1930 to 1953. The College of Agriculture housed the SRS, although its officers came from the Department of Agriculture. At first the Empire Marketing Board gave the department funds for the SRS, but when the board ceased to exist in 1933, the Colonial Office gave the SRS grants under the terms of the Colonial Development Act of 1929 and the Colonial Development and Welfare Act of 1940. Taken together, the college and the Sugar Industry Reserve Fund contributed only about half of the SRS's budget. But during 1934, 1935, and 1936, which were typical years of operation for the SRS, the Reserve Fund committee voted Rs.14,000 for the SRS, while the college contributed Rs.18,000, and the government paid the balance of approximately Rs.34,000.[83] The college was formally a part of the department, and its contribution to the SRS was a budgetary ploy reflecting a consensus in the department and among the planters to favor research over education. But the Reserve Fund's contribution enabled the Chamber to have greater leverage in setting the department's research agenda.

As in the case of the Pamplemousses gardens during the late nineteenth century, the desire of the wealthy producer's organization was dominating a state institution's production and distribution of science and technology. But there were significant differences between the gardens and the department. To some extent, the founding of the SRS followed a trend for British colonial governments to encourage agricultural industries to pay for their own independent commodity research stations, rather than to allow them to rely on government research support. In 1927, India's foremost sugar cane researcher highlighted this approach when he commented that, "The principle . . . of a costly agricultural department being kept up by the Government for the benefit of the European settlers is . . . fundamentally unsound; and the idea is gaining ground all over the tropics that this work should be taken out of the hands of the Government and managed by the planters themselves."[84]

Nevertheless, the SRS was still a government institution whose added funding allowed it to increase the department's volume of sugar cane research. The funding arrangement also invited more meddling from owners of sugar estates, who controlled the Reserve Fund to which all sectors of the sugar industry were required to contribute. Aside from finances, no direct relationship existed between small planters and the SRS. This probably undermined the credibility of sugar cane breeding among the small planters at a time when democratic and anticolonial notions were circulating more widely. If small planters were beginning to question the authority of both the Chamber's members and the undemocratic colonial government, then the Chamber's supervision of sugar cane science was becoming increasingly precarious.

5

SMALL PLANTERS AND NEW CANES, 1913–1937

Some historians working on science and colonialism have drawn a stark contrast between "European" and "indigenous" knowledge. Each culture is widely supposed to construct its own science, with clearly drawn boundaries, while brokers "transfer" knowledge from one culture to another. But when one views science as the systematic production of knowledge, the local sites of production in European colonies begin to seem more complex than the diffusionist model allows. In Mauritius, each individual sugar cane grower produced agronomical knowledge in order to produce the crop. Growers interacted with each other, with the state, and with creditors and debtors. Their knowledge was not uniformly "indigenous," but it was not contingent, atomized, or discordant, either.

In Mauritius, one factor complicated the production and dissemination of knowledge enormously: the persistent scarcity of capital. Money had been in short supply ever since the earliest attempts to colonize Mauritius. In 1909, the Royal Commission pointed to the lack of credit as one of the fundamental problems in Mauritian sugar production. Under Stockdale, the new department of agriculture began to recognize the link between the flow of credit and the availability of plants and training. The government created a department of agriculture that helped the estates, and the government also made loans available to the estate owners. With the small planters, the department implemented joint efforts at extension and cooperation, but by the 1920s it was becoming clear that these were woefully inadequate. The department failed to address the closely related financial and scientific problems of the small planters, which helped to precipitate a crisis in 1937.

Canes and Credit

The small planter cooperative credit movement got off to a promising start, thanks in part to the enthusiasm of the governor in the face of the sugar industry's resistance, and also thanks to a detailed official study of rural credit.[1] In 1913, the governor secured the services of a banking expert from the Indian Civil Service (ICS) who toured the island for several months and made recommendations on how to establish cooperative banks. This ICS officer agreed with the 1909 commission that although lack of credit was not wholly responsible for poor cultivation, "no permanent improvement can be expected without provision of better facilities for credit."[2] He believed firmly in the cooperative banks, which the British had already established in many parts of India. He proselytized among the Mauritian small planters, who had really been hoping for direct government loans and who had no idea of what a cooperative bank was. Nevertheless, he persuaded small planters in several towns and villages to petition the government for cooperatives, before the Department of Agriculture was even firmly established.[3]

In theory, cooperative credit societies strove to improve the lives of their members by encouraging self-sufficiency and thrift and by reducing the influence of village moneylenders.[4] Beginning in 1914, each of fifteen cooperatives had about 120 members, but by the 1930s, as more cooperatives were formed, the average membership dropped to about 65. To become members of a local cooperative, small planters each purchased at least one Rs.10 share, but could not purchase more than 100 shares. This entitled them to apply to the cooperative for a loan. By joining, all members of the cooperative also pledged their property and their unlimited liability for any member's default on a loan. This collateral encouraged confidence in the cooperatives, because it created an incentive for mutual surveillance. Each cooperative had a board of about eight directors who were chosen by the members and approved by the Department of Agriculture's inspectors. They were responsible for deciding which members deserved loans, as well as for keeping the cooperative's accounts in order. The cooperatives also raised capital by encouraging members and non-members to deposit their savings. They also received small government loans at low interest.

The cooperatives served mainly to promote the interests of the wealthier small planters, despite their lofty goals and the ceiling of 100 shares per member. Numerous complaints surfaced about wealthy members who abused their privileges.[5] Cooperatives also excluded sharecroppers and tenant farmers, because their regulations required members to pledge land as security. At the same time, common estate laborers earning Rs.30 per month during the harvest would have found the Rs.10 share somewhat beyond their means.[6]

Despite the existence of the cooperatives, the system of crop finance did not change for the majority of small planters. They continued to borrow from family members and moneylenders, even if the cooperatives charged lower rates of interest. Many preferred not to have their financial matters discussed in public before the cooperative's board of directors, for fear of attracting creditors and tax collectors. Small planters' title deeds to lands were sometimes not in order either, and moneylenders did not require detailed proof of ownership. Furthermore, moneylenders could often provide capital on shorter notice than the cooperatives.[7] In some cases, moneylenders lowered their interest rates to compete with new cooperatives.[8] And during the boom years of the early 1920s, estates offerred easier credit to small planters, diminishing the need for cooperatives except in remote areas.[9]

The cooperatives had only a limited economic impact. Between 1914 and 1937, their total membership never exceeded 3,360, and it averaged 2,639 for the nineteen years for which figures are available, about 4 percent of the male population engaged in agriculture.[10] (See Table 2.) The total loans that the cooperatives granted amounted to between 2 and 4 percent of the small planters' estimated earnings from sugar in most years, except for the boom year of 1921, when it reached a maximum of 5.3 percent.

The cooperatives had a minimal financial impact, but they were of considerable importance when it came to the dissemination of agricultural science and technology. This was because government agricultural extension officers supervised them. Before the cooperatives, creditors made certain that small planters who borrowed money were reasonably good farmers. But the new Department of Agriculture sponsored the cooperatives and applied different standards. The first government inspector of cooperatives came from India on a temporary appointment. He was an expert on cooperation, and he helped to found the original 23 cooperatives before he left the island in 1915. After that point, agricultural extension officers served as inspectors of cooperatives until the creation of a separate Department of Cooperation in 1947. On the one hand, the government appointed agricultural extension officers to inspect cooperatives because it believed

Table 2

The Impact of Cooperative Credit Societies in Mauritius, 1914-1937

Year Ending June 30	Number of Societies	Number of Members	Number of Members Who Borrowed	Total Amount Loaned (in Rupees)	Estimated Sugar Income of Small Planters*
1914	15	—	—	115,504	10,370,213
1915	20	—	—	182,015	8,339,250
1916	23	—	—	—	8,564,844
1917	23	—	—	245,929	8,487,204
1918	23	2767	—	262,636	10,172,350
1919	23	2781	1746	255,312	17,214,497
1920	26	3072	1814	350,244	43,461,698
1921	33	3504	1511	412,673	7,850,406
1922	34	3641	1555	376,336	10,574,168
1923	36	3660	1694	320,579	12,730,301
1924	32	3209	1523	220,814	9,269,288
1925	—	—	1606	201,138	7,434,883
1926	30	2994	—	—	7,391,412
1927	29	2861	1589	165,701	6,913,434
1928	26	2492	1507	164,044	7,137,983
1929	26	2415	1530	133,386	5,910,880
1930	27	2365	1550	156,070	4,528,133
1931	29	2350	1618	118,101	3,255,516
1932	28	2252	1553	99,066	5,054,041
1933	28	2101	1489	141,496	5,220,049
1934	28	1865	1184	100,728	3,576,842
1935	29	1902	1221	107,175	5,276,957
1936	29	1882	1110	117,449	5,414,264
1937	34	2032	1160	177,536	6,289,435

* Note: this estimate is intended only to give a rough idea of the capital circulating in the small planter sector. It was arrived at by taking the average annual net producer price of sugar and multiplying it by the island's annual sugar production. Twenty-five percent of this figure was taken, representing an estimate of the average sucrose production of small planters. This figure varied over the years, as small planters cultivated more or less lands under cane depending upon the price of sugar. Their fields yielded lower tonnage of canes than estates, however, with even lower tonnages in marginal lands placed under cane in boom years. The sucrose content of their canes was probably also less than estate canes. Two-thirds of the figure was then taken, because factories typically retained one-third of small planters' sugar as a processing fee.

Sources: *Reports on the Working of Cooperative Credit Societies in Mauritius,* 1914-1947. Production figures and prices taken from The Mauritius Chamber of Agriculture, *The Mauritius Chamber of Agriculture, 1853-1953* (Port Louis: General Printing and Stationery Co., 1953).

that they could assess cultivation practices better than specialists in banking, and they were already spending a considerable time traveling through the countryside. On the other hand, the extension officers could use the cooperatives as centers for the distribution of information.

As a result, a small number of wealthier small planters who managed the cooperatives obtained the research results of the Department of Agriculture before the majority of poorer planters. In the short run, this gave wealthier small planters leverage in their credit relations with poorer smaller planters because they could criticize poorer planters' plots based upon knowledge garnered from the state. In the long term, poorer planters learned about scientific cultivation practices, but not until the wealthier planters had taken their pound of flesh in the form of higher interest payments. Like the estates, which permitted the Department of Agriculture to establish experimental plots on their lands, members of cooperatives who allowed extension officers to plant new crops on their plots gained a small advantage in knowledge and productivity over their neighbors.

Extension officers lost a great deal of time by depending on the cooperatives, diminishing the overall effectiveness of the Department of Agriculture with small planters. While they did use the cooperatives to organize demonstration plots for new food crops and cane varieties, supervising the financial operation of the cooperatives was a tedious and time-consuming affair that detracted from their purely agricultural work. During the 1910s and early 1920s, the department had only one full-time agricultural instructor for the entire island. In 1916, he reported that the cooperatives took up "much" of his time, and by 1923 they accounted for "most" of his time.[11] During the late 1920s and 1930s, the Department of Agriculture added two additional extension officers to its staff, but this was still insufficient to address the needs of approximately 30,000 small planters. Two of the three officers inspected cooperatives as part of their duties, even with the addition of a full-time assistant registrar of cooperatives in 1933 and a full-time inspector in 1935.

The handful of extension officers were bogged down in a multiplicity of tasks and lacked time for instruction in the cane fields. Aside from overseeing the work of the cooperatives, they supervised a government land settlement scheme, begun in 1934 as a way of relieving unemployment.[12] They also gave instruction in school gardens, and between 1928 and 1930 some taught in an agricultural secondary school. However, all efforts to combine education with farming proved unpopular among both teachers and students because most of them had attended school in the hope of never having to farm again. The extension officers were also responsible for instruction in crops other than sugar cane. They tended to spend much of their time teaching about food crops, because the government sought to increase the self-sufficiency of the island.[13] The officers left the responsibility of teaching small-scale cane planters to the estates and the sirdars, because of understaffing and estate resistance. Transportation problems compounded the lack of extension staff. The government could not afford to supply automobiles to extension officers, and so they relied on a time-consuming combination of railroads and bicycles to visit farmers.

Extension services also suffered from poor planning and restricted government funding. Following the recommendations of the 1909 commission, Frank Stockdale blamed the "faulty cultivation" of small planters on ignorance, lack of capital, and "the temperament of the Indian." He hoped that a combination of cooperative credit and extension would improve the small planters' cultivation methods, raise the sugar output of the colony, and thereby increase the government's revenues. He wanted to use itinerant agricultural instructors to achieve these goals, following the model of the British West Indies. Failing that, he thought the cooperatives might be used to demonstrate agricultural techniques as they had in the Punjab. But he staffed his program with only one instructor, far short of the island's requirements.[14] Perhaps his colonial bias led him to believe that local people were not capable of higher standards of cultivation, an opinion shared with agricultural officers in other colonies.[15] Some estates opposed extension services, on the grounds that extension and education might reduce laborers' interest in estate work, and they may have persuaded the director to moderate the directives of the 1909 commission.[16] In any case, he set a precedent of understaffing that lasted until 1937.

The 1909 commission blamed the small planters' "yield gap" on lack of credit and lack of knowledge, but the Department of Agriculture's cooperatives and extension services failed to address these related problems in an adequate fashion. In 1929, when the Colonial Office sent an expert to investigate the Mauritian sugar industry, he found it necessary to remind the department of the importance of agricultural education for the small planters.[17] Stockdale, now agricultural advisor to the Colonial Office, responded that either the Mauritian government ought to improve Indo-Mauritian cultivation methods or they should work to get the land back into the hands of the large estates.[18]

In the early 1930s, the department made an effort to heed Stockdale's recommendations about improving small planter cultivation. Extension officers created several more demonstration plots, but staffing was still insufficient, and research officers never visited small planters' fields. But still, the yield gap remained quite wide. Between 1920 and 1937, estates yielded on average 51.9 metric tons of cane per hectare, more than double the small planters' average of 25.9 tons. (See Table 3.)

And yet, one junior officer showed the potential for extension services to blend the Mauritian scientific tradition with recognized academic training and to deliver the results to planters. His name was Alfred North-Coombes, a man who would become the most famous agriculturist in twentieth-century Mauritius. Like his supervisor, Charlie O'Connor, North-Coombes was descended from a British official who married into a Franco-Mauritian family. Much of the North-Coombes family was involved in the Mauritian sugar industry, and Alfred himself was born on the Britannia sugar estate where his father was the manager. But unlike O'Connor, North-Coombes had formal academic training in agriculture. In 1928, young North-Coombes graduated first in his class from the College of Agriculture and won a scholarship to the University of Reading, where he earned a B.Sc. in Agriculture in 1931. After a brief stint working on a farm in Denmark, he was appointed as a lecturer in the Mauritius College of Agriculture and as an agricultural instructor in the Department of Agriculture.

North-Coombes spent most of his time in extension work and other departmental duties, but he was unusual in that he was equally at home working with small planters, estate staff, and government officials. In his spare time, he was making a reputation for himself as one of the island's leading intellectuals. In 1937, he published a history of the Mauritian sugar industry which remained the definitive treatment until 1993, when he published a second edition. Over the course of a long career in the department, he became an authority in Mauritian agriculture, biology, and history, a polymath in the tradition of Poivre, with the training of Tempany, and with fluency in English, French, and Kreol.

But even North-Coombes believed that his broad responsibilities stretched the limits of his knowledge, and in an interview many years later he admitted frankly that many planters knew more than he did about some of the crops and techniques he was promoting.[19] During the 1930s, as the sugar economy worsened, Mauritian small planters needed more help than North-Coombes and his overworked colleagues could deliver.

Agriculture during the Depression

During the 1920s, most small planters lacked direct access to the colonial department of agriculture. At the same time, Indo-Mauritians and Creoles

Table 3

**Metric Tons of Cane Harvested per Hectare:
Estates with Factories and Small Planters, 1920-1952**

Year	Estates with Factories	Small Planters	Island Average
1920	56.2	26.4	38.1
1921	46.4	20.7	31.7
1922	46.4	28.6	35.7
1923	47.8	19.0	31.7
1924	52.1	22.6	36.9
1925	55.5	23.3	39.5
1926	47.1	21.7	33.6
1927	50.2	19.0	37.1
1928	55.0	24.3	42.1
1929	50.5	30.5	40.5
1930	48.6	37.1	45.5
1931	37.1	23.6	32.4
1932	55.7	29.0	46.4
1933	57.4	27.8	47.4
1934	43.3	16.2	33.1
1935	59.3	31.4	50.2
1936	60.5	30.2	49.7
1937	64.7	34.0	55.0
1938	61.6	28.1	51.4
1939	48.3	24.8	42.4
1940	61.4	32.6	52.6
1941	60.0	29.3	51.6
1942	59.3	29.5	54.0
1943	63.1	35.5	54.7
1944	47.1	26.9	40.0
1945	33.6	20.2	28.3
1946	60.2	35.2	49.7
1947	64.5	28.1	50.9
1948	70.0	31.2	55.2
1949	70.7	34.7	55.5
1950	73.5	37.4	58.5
1951	78.5	45.5	65.0
1952	73.3	41.4	59.0

Source: The Mauritius Chamber of Agriculture, *The Mauritius Chamber of Agriculture, 1853-1953* (Port Louis: General Printing and Stationery Co., 1953).

continued to agitate for government reform, even though they did so separately and achieved very little. Indo-Mauritian intellectuals drew inspiration from their Indian roots, while Creoles celebrated their French heritage and often denied any African identity. The growing Indo-Mauritian elite of merchants and professionals looked to the example of the Indian National Congress and took up the cause of the Indo-Mauritian community. For their own part, some Creoles launched a campaign to return Mauritius to French rule.[20]

The economic troubles of the 1920s and 1930s exacerbated economic, social, and political tensions. Between 1914 and 1923, prices were generally high, especially during the boom year of 1920 when they rose to 1013.60 rupees per metric ton. Workers could earn three rupees per day, and they began to eat better, work less, and even pursue leisure activities. But by the mid-1920s, sugar prices had begun to decline. They fell steadily until they reached an all-time low of Rs.109.90 per metric ton in 1936. Daily wages during the cane harvest dropped to sixty cents for men and thirty cents for women, so that even regular workers went hungry.[21] The system of Imperial Preference did not alleviate the depression, even when the International Sugar Agreement of 1937 stabilized prices and gave Mauritius a quota of 267,000 metric tons on the British market.

The depression of the 1930s struck the island hard and helped a new class politics to eclipse identity politics, at least for a time. The first indications of working-class unrest came during the mid-1930s. The first Mauritian to take up the estate workers' cause in earnest was Dr. Maurice Curé, who had experienced workers' problems firsthand through his medical practice. He advocated a broader franchise, judicial reform, and subsidised loans to small planters, as well as a socialist package of better wages, housing, health care, and the right to form trade unions. During the late 1920s and early 1930s, he gained a reputation as an agitator for reform. In 1936, he founded the Mauritius Labour Party, as well as a workers' organization called the Société de Bienfaisance des Travailleurs. He and his assistants recruited members throughout the island, proving the possibility of cross-cultural cooperation. During late 1936 and early 1937, both Creole and Indo-Mauritian workers joined by the thousands. Curé also put pressure on the local government by keeping in contact with the Fabian Colonial Bureau and the British Labour Party in London. The governor refused to cooperate with the Mauritius Labour Party, judging it to be unrepresentative. The Colonial Office explored the possibility of conciliation and compromise with Curé's organization, but British administrators "on the spot" contemplated using force against it.[22]

The colonial government responded to the post-1923 decline in prices by assisting the sugar estates, but the government did little to help other producers. In 1925, the government reduced export taxes on sugar, a measure that increased the entire population's relative tax burden. In 1926, 1929, and 1931, the government provided large, long-term loans for the persistently capital-starved sugar industry, to the point of straining the colony's budget. The Department of Agriculture also collaborated with the Chamber of Agriculture on experiments in field mechanization, so that estates could reduce their wage bill. Small planters still received little in the way of extension services or loans from cooperative credit societies, while workers did not benefit at all from government largesse.[23]

Between 1929 and 1932, as the sugar industry faced a worsening crisis, the Department of Agriculture passed through a period of transition. In January 1929, the Sugarcane Research Station (SRS) began work when its senior geneticist and plant breeding officer, Glendon Hill, transferred to Mauritius from Nigeria. But in February, just as the SRS was getting under way, the Colonial Office appointed Tempany to be the new director of agriculture in Malaya. In some ways, his departure resembled Stockdale's. His relations with the sugar estates were becoming difficult, but he also seems to have been in line to hold greater responsibilities. Curiously, the Colonial Office did not appoint someone with equivalent credentials to replace him. Instead, they named the department's elderly entomologist, Donald d'Emmerez de Charmoy, to the directorship.

His appointment came a time when there was some bad blood between the Chamber and the Department of Agriculture. D'Emmerez de Charmoy was an entomologist, and the Colonial Office may have appointed him as a way to signal its interest in solving the *Clemora* problem. But his appointment may also have been a gesture of conciliation. D'Emmerez de Charmoy had been serving as Tempany's assistant director, but he was also a Franco-Mauritian from a family with a long-standing involvement in the sugar industry. He was also a scientist in the local tradition. In any event, he was only director for a short time: he died in November 1930. The Colonial Office named Glendon Hill to be acting director because he was the next-highest ranking civil servant, even though he was supposed to be working full-time as a sugar cane breeder.

It was not until 1932 that the Colonial Office named a new director, Gilbert Bodkin. This was Bodkin's first appointment as a director of a department of agriculture, but he brought with him a broad range of experiences in the colonial agricultural service. He earned his B.A. at Cambridge in 1908 and stayed on to study for the diploma in agriculture, which he received in 1910. He served as a government entomologist in British Guiana from 1911 to 1922 and in Palestine from 1922 to 1932. His background in entomology suited him well to the directorship in Mauritius, because of the sugar industry's problems with *Clemora*. Nevertheless, during his service in Mauritius, which lasted from 1932 to 1948, he put most of his energy into departmental administration.[24]

Bodkin maintained the established trajectories of the department's work. Agronomists and chemists continued their basic research in cultivation practices. Tractors virtually disappeared during the Depression, but agronomical studies of fertilizers and soil increased the estates' efficiency at a time when pressure not to waste capital on unnecessary fertilizers was at its height. Still, in 1934 the Society of Chemists, made up of estate and factory staff members, urged the department to expand its fertilizer trials on estates.[25] Most importantly, beginning in 1936 the department introduced the technique of "foliar diagnosis," the chemical analysis of cane leaves. Previous methods involving the analysis of cane juice had not revealed deficiencies in soils nearly as well.[26]

The department's agronomists and chemists were making steady progress, but the entomologists and pathologists were still failing in their efforts to eradicate pests and diseases, which increased pressure on agronomists to improve cultivation and on cane breeders to produce new varieties. During the early 1930s, the *Clemora*, gummosis, leaf scald, and red rot were all attacking the Tanna canes. In 1931, the SRS director noted that it was only the "gradually improving standard of cultivation on the estates" that maintained cane yields, "in spite of the steady spread of Phytalus [*Clemora*], and the acknowledged deterioration of the White Tanna cane."[27] The beetle infested 42 percent of all fields under cane, while the island's average cane sucrose content had dropped from about 13.3 percent in 1915 to 12.7 percent in 1930.[28] During the same period, estates planted Tanna canes on approximately the same percentage of lands under cane: Tannas occupied 56 percent in 1915, reached a peak of 63 percent in 1925, and fell to 57 percent in 1930.[29] By this point in time, virtually every other major cane-growing region in the world had replaced its old noble varieties with seedling canes.

Still, the continuing onslaught of the *Clemora* made the Mauritian department pessimistic. For twenty years they had failed to control this insect and failed to breed a cane to replace the Tannas. In 1932 Bodkin lamented, "What of the future? Will this destructive insect forever continue to deprive the sugar industry of this island of a substantial part of its hardly earned profits?" Looking to the root investigations, he remarked "that some grounds most certainly exist for hoping that the reign of this tyrant insect will in due course come to an end."[30]

The Hybrid Revolution

The 1930s were bleak in many respects, but they were the golden years of plant breeding. Breeders were starting to reap the benefits of thirty years of basic genetic research. Before 1900, scientists had come to accept Darwin's theories of evolution and natural selection, but nobody knew what principles governed variation and heredity. Nobody, that is, except Gregor Mendel, but his work on particulate inheritance was neglected until European scientists "rediscovered" it around 1900.

Even though this rediscovery came late, it revolutionized the life sciences and it provided fundamental concepts for the new science of genetics. In 1909, Wilhelm Johanssen established the difference between genotype, which is the genetic constitution of an individual, and phenotype, which includes all observable traits, both genetic and environmental. Johanssen's distinction between genotype and phenotype guided breeders who wished to distinguish inherited characteristics from characteristics that were influenced by external conditions. Plant breeders learned that they needed to isolate genotypes from phenotypes if they wished to select for hereditary characteristics. At the same time, new mathematical methods such as regression analysis made it easier for breeders to analyze and predict the results of hybrid crosses.[31]

During the 1910s and 1920s, scientists still knew very little about the genetics of the sugar cane plant. Breeders were still relying on the empirical methods that Morris and Stockdale described in 1906: they combined the largest possible number of varieties and then evaluated how well the seedlings performed. Out of the thousands of cane seedlings growing in the first round of the trials, only a handful would be worth a second trial, and very few would ever grow on sugar estates. Johanssen's distinction between genotype and phenotype was still not widely applied to sugar cane. With hindsight, the selection of varieties based solely on weight and appearance seems almost futile. Genotypes had to be isolated. As G.C. Stevenson wrote in his 1965 textbook on sugar cane breeding, "The success of a breeding programme will depend not only on the frequency with which such new combinations result in individuals of improved commercial worth, but also on the methods which are used to detect them in large populations, obscured as they often are by unproductive environmental variation."[32]

Starting in the 1920s, breeding methods became more sophisticated in most sugar cane research stations, and the Dutch laboratory in Java, POJ, was the most sophisticated of all. The Dutch had attempted hybrid crossing previously but obtained only disappointing results. Those first crosses between noble canes (*S. officinarum*) and wild canes (*S. spontaneum*) yielded progeny that had the hardiness and disease-resistance of wild canes, but the progeny did not yield as much sucrose as the pure noble varieties. But during the 1910s, Jakob Jeswiet and his colleagues at POJ began to make systematic hybrid crosses in which they consciously distinguished genotypic characteristics. They learned that the most productive hybrid canes derived from "backcrossing" noble canes with wild canes. First, they crossed a noble cane with a wild cane, yielding a noble-wild hybrid; next they crossed the noble-wild hybrid with another noble cane, yielding progeny that was one-fourth wild; then they crossed the one-fourth wild cane with another noble cane, yielding progeny that was one-eighth wild. Jeswiet called this process "nobilization" because the hybrid canes were gradually becoming more "noble," yielding more sucrose while still retaining as much of the wild cane's disease-resistance as possible. (He could use hybrids as parents in crosses because sugar cane hybrids are usually fertile, unlike hybrids of many other animal and plant species.)

In 1921, Jeswiet's nobilization program yielded the most famous hybrid cane of all, POJ2878. He began his experiments in 1914, with a cross between the noble seedling cane POJ100 and a cane called Kassoer, a naturally occurring hybrid of *S. officinarum* and *S. spontaneum* found growing wild on the slopes of a volcano. By 1917, Jeswiet was able to select the most promising one-quarter-wild progeny, POJ 2364, and he crossed it with the noble seedling cane, EK28, which itself was the product of a cross between POJ100 and another noble seedling cane, EK2. By 1921 the cross of POJ2364 with EK28 had yielded a third generation of canes that were one-eighth wild. They contained sufficient sucrose but had remarkable disease-resistance. The best product of the cross between POJ2364 and EK28 was POJ2878, a cane that not only saved the Javan sugar industry from the sereh disease, but which the Dutch scientists exported to the rest of the world.[33]

It is interesting to note that sugar cane researchers exchanged plants and information freely with each other, regardless of nationality. Jeswiet described his nobilization methods to fellow cane breeders attending the 1927 meeting of the International Society of Sugar Cane Technologists.[34] There may be practical reasons why breeders did not compete. By the early twentieth, marketing agreements had diminished economic competition between cane-growing regions, meaning that there were no national or imperial barriers to the international transfer of cane science and technology.[35] But there were other reasons for early twentieth-century cane scientists to swap specimens and ideas amongst each other so freely. As Robert Kohler argues in his history of *Drosophila* experimentation, professional codes attached moral significance to cooperation among the producers of biological knowledge.[36] Besides, had there not been cooperation, it would have been easy to pilfer new canes from a researcher's fields.

At the same time as POJ was breeding hybrids, the station was also doing the pioneering research on cane cytology, the study of the cane's chromosomes. Biologists had only formulated the chromosomal theory of inheritance during the 1880s. Walther Flemming discovered chromosomes in the nuclei of dividing salamander cells, and August Weissman observed that each gamete (sperm or egg) contains half the number of chromosomes as each somatic cell's nucleus. According to Weisman's theory, the number of chromosomes is halved at meiosis and is restored after a sperm fertilizes an egg. Therefore, half of an organism's genetic material belongs to the father and half belongs to the mother. During the 1910s, T.H. Morgan's research on *Drosophila* showed that some traits were inherited from one particular parent; that these traits were based on particular genes; and that the positions of genes could be "mapped" on chromosomes.[37]

During the 1920s, POJ's G. Bremer used Morgan's insights to study the chromosomes of the sugar cane plant. All canes have diploid cells, meaning that they have two chromosomes of each type. All true noble cane cells have 80 chromosomes, meaning that their gametes have 40 chromosomes. The cells of the *S. spontaneum* wild varieties can have between 40 and 128, and their gametes contain half that number, too. Theoretically, then, it is possible to predict the number of chromosomes in the cell of a hybrid cross between a noble cane and a wild cane: the number should be 40 plus half the number of chromosomes in the wild cane. When Bremer used wild canes as female parents and noble canes as male parents, this turned out to be the result. But when Bremer used noble canes as female parents and wild canes as male parents, he learned something surprising: the hybrid progeny had 80 chromosomes, the full complement of noble chromosomes, plus half the number of chromosomes in the wild cane. Bremer's insight had profound implications for backcrossing: when breeders used noble females as parents, there would be some restitution toward the female side. In other words, the phenotype, or observable characteristics of the hybrid cross, would tend to resemble the female. But Bremer also learned that this process of phenotypic restitution only occurred for two generations and not for the third.[38]

By the end of the 1920s, POJ's results were pushing scientists in other sugar cane experiment stations to the conclusion that noble canes had a limited gene pool. Better canes could be obtained by introducing genetic material from the other species of *Saccharum*. Hybridizing experiments became more extensive, and several stations made especially noteworthy contributions. In Barbados, the new British West Indies Central Sugar Cane Breeding Station refined many of the techniques for hybridization. At the Coimbatore station in India, Barber and Venkatraman created some of the most important hybrids for subtropical cultivation.

Scientists in Barbados, India, and Hawaii also learned a number of practical things about sugar canes, such as the fact that male-sterile varieties could be used safely as female parents, eliminating the need to remove anthers from flowers. Different research stations developed different techniques for isolating the pollen of flowering canes, thereby making it possible to be sure which cane was the male parent. Barbadian breeders developed the "lantern," a covering that protected cane arrows from random fertilization. Indian researchers isolated flowering canes by cutting them down and replanting them, and they also studied ways to get parent canes to flower at the same time. In Hawaii, researchers learned that they could eliminate unwanted pollen by using a mild solution of sulphuric acid. All these techniques came into use in the world's sugar cane breeding stations. But even with new techniques and new knowledge, the breeding of canes remained a time-consuming and labor-intensive process. A cane breeder had to plant tens of thousands of new seedling varieties in order to begin a sequence of trials that might last up to ten years, and which in the end only had the potential to produce one or two useful new varieties.[39]

POJ's work on sugar cane hybridization had a profound effect on the breeding program in Mauritius. During the early 1930s, the *Clemora* was destroying more and more of the Mauritian sugar crop, while world sugar prices were plummeting. It became the official mission of the SRS to produce "vigorous hybrid seedlings of high-yielding and disease-resisting powers to replace those varieties in the island which now prove inefficient."[40] The SRS applied the new methods of plant breeding, with the result that colonial agricultural scientists in Mauritius finally turned the corner and produced the first in a series of hybrid "wonder canes" that transcended local agricultural problems.

During the 1930s, some of the most important cane breeding experiments were conducted under the supervision of Glendon Hill, who directed the SRS from 1929 to 1937. After he departed to direct the Amani research station in Tanganyika, the SRS was placed under the chemist,

Norman Craig. The SRS hired a new senior geneticist, G.C. Stevenson, a Cambridge B.A. with a diploma from the new Imperial College for Tropical Agriculture in Trinidad and with experience breeding sugar canes in Barbados. Stevenson was an intense, academic type who later went on to publish the definitive textbook on sugar cane breeding.

But while Hill and Stevenson headed plant breeding at the SRS during the 1930s, a group of Mauritian scientists conducted the selection trials. They were Aimé de Sornay, the Cane Breeding Officer, and his assistants, G. Mazery, André d'Emmerez de Charmoy, and Pierre Halais, all of whom were graduates of the Mauritius College of Agriculture. Of them, the career of Aimé de Sornay had the greatest impact on the Mauritian sugar industry. His work also demonstrates some of the cultural complexities of colonial science in Mauritius. He graduated from the Mauritius College of Agriculture in 1928, but instead of doing postgraduate work in England, as North-Coombes had done, he spent a year in Paris studying at the Institut d'Agronomie Coloniale. He was from a family that was rooted in the Mauritian scientific tradition, but like Moutia, North-Coombes, and Vinson, he received his formative training at the College of Agriculture. And he was interesting, for a scientist in a British colony, in that he turned to France for further training.[41]

In 1929, Aimé de Sornay went to work for the SRS. Trials of the newly imported varieties proved once again that it was better for a regional industry to breed its own canes. The Chamber had promoted POJ2878 and other foreign varieties out of frustration with the department, but none of the imports met the combined criteria for high yields and disease resistance that the SRS established. Some of the imports did yield more highly than the Tannas, such as POJ2878, which the president of the Chamber referred to hopefully as the "queen of canes."[42] It resisted most of the important local diseases too, but its susceptibility to leaf scald and its poor factory qualities relegated it to a minor proportion of the cane area planted. However, its good characteristics meant that it could play a major role as a parent cane in the new Mauritian breeding program.[43]

Aimé de Sornay's first task was to study the characteristics of sugar canes already present in Mauritius. As he indicated in an article published in 1930, he was following the examples of POJ and Coimbatore. He made observations on the botany of different cane flowers, paying special attention to the fertility and longevity of different types of cane pollen. He tested different methods for crossing canes and compared the male and female fertility of cane varieties. He also studied the rooting of seedlings and compared methods of fertilizing them.[44] Aimé de Sornay was responsible for providing fundamental knowledge for the new breeding program at the SRS, and he published his findings in French, the language of the sugar estates, rather than in English, the language of his supervisors. He may have done this because the estates were important for the breeders. Depending on the year, between eight and ten of them let the department borrow land for the later stages of varietal trials.

From 1930 to 1932, Aimé de Sornay embarked on a program of hybridizing and nobilizing canes, following the example of Jeswiet. Aimé de Sornay's results confirmed Jeswiet's methods: nobilized cane varieties in the generations after the original hybrid cross produced the best results. But Aimé de Sornay worked out an important difference between Java and Mauritius. Jeswiet's best cane was POJ2878, a third-generation nobilization that was seven-eighths noble and one-eighth wild. At the instigation of the Chamber of Agriculture, Tempany had imported this cane. But trials soon showed that POJ2878 did not grow well in Mauritius. In fact, Aimé de Sornay's trials were indicating that fourth-generation nobilized canes with one-sixteenth wild pedigrees were better suited to Mauritian conditions.

Aimé de Sornay applied Jeswiet's methods to nobilize canes, and he also used his knowledge of genetics. The interdisciplinary nature of the SRS helped enormously in this regard. During the early 1930s, the SRS botanist, Harry Evans, investigated cane root systems and showed that certain canes, particularly the hardy canes descended from the noble crosses with *S. spontaneum*, resisted *Clemora* as well as the island's other pests and diseases. Evans's research provided breeders with an important insight: they needed to select canes with hardy root systems.[45]

In the face of the continuing *Clemora* threat, the SRS turned to breeders to produce a cane with a root system that was sufficiently strong to tolerate the destructive larvae. This in itself was a significant departure for cane breeders: previously, they selected canes based on visible growth habits and sucrose yields. But if a strong root system could help to resist *Clemora*, then it, too, had to become a criterion for selection.

Aimé de Sornay had to select for this additional characteristic. To ensure accuracy, he combined his tacit experience of sugar cane selection with his new knowledge of sugar cane genetics. By this time, the actual method of mating two canes was fairly stable. The breeders cut down the female cane and attached it to a stalk of bamboo that was stuck in the ground. Next they cut down a flowering male cane and stood it in a mild solution of sulphuric acid. They placed the male cane next to the female cane and enclosed them in a linen bag for twenty days. The resulting seeds were germinated and planted. In 1953, Aimé de Sornay wrote an article (in

French) that reflected back on the work of the early 1930s. He stated that the crossing of canes was relatively easy; it was the selection of seedlings that was difficult. According to him, "The geneticist must have flair and experience in order to recognize the seedlings that offer the most promise. He must be able to differentiate between genetic variation and the kinds of variation that are due to extrinsic factors." He was beginning to distinguish the cane's genotype from its phenotype, but much depended on "flair." He recognized that much research would have to be done in order to get a full understanding of the cane's complex genetic composition.[46]

The expanded breeding program at the SRS, together with ongoing work in agronomy, botany, and chemistry, soon paid huge dividends for the sugar industry. During the standard four years of testing new varieties, Aimé de Sornay and his SRS colleagues found that even some of its earliest hybrids raised in 1930 and 1931 performed better than the Tannas and the other imports. By 1936, the department was propagating twenty varieties bred in Mauritius and reported increasing applications for cuttings from growers.[47]

Ever since the 1880s when Mauritians recognized the possibility of producing seedlings, local scientists had been searching for a wonder cane. This philosopher's stone of the sugar industry would have high yields, grow everywhere on the island, and resist diseases and pests. In 1936, the SRS began to realize that it had such a cane in its seedling collection: the mundane-sounding M134/32, which Aimé de Sornay and his team of Mauritians created by crossing POJ2878 with D109, a seedling cane imported in 1905 from British Guiana. Finally, the government would have a seedling cane, a fourth-generation nobilized hybrid, even, with the potential to save the sugar industry. But just as a new era of cane breeding was dawning in Mauritius, events in a remote corner of the island called attention to some further problems in the production and distribution of colonial sugar canes.

The Uba Riots of 1937

The demise of the Tanna canes and the global economic depression caused the Chamber to apply pressure to the Department of Agriculture and the SRS to provide new canes. Thanks to the new genetic approach to cane breeding, to the interdisciplinary collaboration at the SRS, and to the efforts of Franco-Mauritian scientists trained and employed by the Department of Agriculture, the sugar estates did get better canes. But the sugar barons were not the only farmers to suffer from declining varieties and collapsing markets. The small planters also suffered, and they, too, looked to new cane varieties for salvation. The depression of the 1930s provoked disorder in some of Britain's sugar colonies. In Mauritius, access to sugar canes was a central grievance of protesters. Between 1913 and 1937, the small planters were faced with some of the same problems as the estates, but the state did little to help them. The price of sugar declined steadily, except during the unusual years of 1914–1921, and the chronic lack of finance capital squeezed all producers. By 1936, when the price of sugar hit rock bottom, the small planters merited more government attention than ever.

Like the estate owners, the small planters turned to new breeds of sugar cane in order to increase their profit margins. However, given the negligible effect of the extension services over the years, they could not hope for much state assistance. Instead, the small planters began to identify and plant new varieties of their own choosing. Most of these were known locally as "Uba" canes. They grew well in adverse conditions, produced heavier cane yields than all other varieties, and thereby narrowed the gap with estate-grown canes. Factories paid small planters according to the weight of the canes and disliked the fact that Uba canes had low sucrose content. The small planters used Uba canes to increase their receipts, while the factories that processed these canes had less sugar to sell for every rupee they paid to the small planters.

Ironically for the elite, they could only blame their own agricultural and scientific initiatives for the introduction of the hardy, low-sucrose Uba canes. During the bad years of the late 1860s, the Chamber of Agriculture noted how sugar producers in Natal were cultivating a variety called the "China cane," and requested the Pamplemousses gardens to acquire some cuttings.[48] The gardens obliged, and a rain-soaked box of cuttings arrived shortly thereafter; it came via Durban, but only the letters u-b-a were legible on the label, hence the name Uba. These canes were actually a different species of cane, now classified as *Saccharum sinense*.[49] Trials at the gardens probably revealed both their vigorous growth and their low sucrose content, and although Uba canes were listed in the the gardens' catalogue, they were not cultivated on a commercial scale in Mauritius.[50]

And yet, the Uba canes continued to play a marginal role in the production of knowledge about canes. Early in the decade of the 1900s, when the Chamber began to feel frustration with the Station Agronomique's lack of success in cane breeding, one member did float a proposal for cultivating the Uba cane.[51] In 1914, the Department of Agriculture even used it in some of its earliest varietal trials.[52] Despite these efforts, the Mauritian estates never grew Uba. But the cane's hardiness and disease resistance made it popular at various times in Natal, Louisiana, Puerto Rico, and other places.[53] Factory owners in Natal adjusted their crushing machinery to Uba's higher vascular fiber content in order to compensate for the cane's lower sucrose content.[54] The Mauritian estates had a climate favorable to a broader range of varieties and preferred simply not to cultivate the Uba rather than change their factory methods.

During the 1920s, some estates experimented with several other new varieties called Uba, although none actually had the original Uba in its pedigree.[55] Several Mauritian estate-owners made or discovered natural hybrid crosses between noble canes and a variety of the wild *S. spontaneum* that Hindu immigrants imported during the nineteenth century to plant near temples and to use in religious ceremonies. Therefore it is possible that small planters interpreted the appearance of Uba on the landscape differently from the estates and the Department of Agriculture, who tended simply to emphasize crop yields, sucrose content, and disease resistance. The estates called the new hybrids Uba because they resembled *S. sinense* in their vegetative vigor, high fiber, and low sucrose. One of these varieties was called Uba Marot, after the owner of an estate in the Black River district who found it growing in his fields.

Most of the estates that attempted to cultivate the Uba cane rejected it on account of its high fiber and low sucrose content. Uba canes produced only 74 percent of the sucrose one could expect from a standard noble variety. The Department of Agriculture confirmed these overall results, but found that Uba Marot might be more useful if it were manured heavily and allowed to mature longer than most canes.[56] The manager of the "Rich Fund" estate in the Flacq district also found similar seedlings of a cross between *S. officinarum* and *S. spontaneum* growing in his fields. The estate named these canes Uba de Rich Fund, and after planting them on up to 20 hectares during 1926 and 1927, it found them to have great vigor but too little sugar for the factory to make a profit. Nonetheless, up until 1937 the estate continued to cultivate several hectares of Uba de Rich Fund because it grew well on marginal lands.[57]

Uba Marot and Uba de Rich Fund resulted from the kind of selection trials that Mauritian estate owners had been conducting on their own for some time. But it was Indo-Mauritian small planters who recognized that these new varieties represented significant new opportunities. Small planters in the vicinity of the two estates acquired cuttings of the Uba canes, and found them ideally suited to their needs in a difficult time. In Flacq, a small planter who acted as a cane broker between the estates and other small planters recognized the Uba cane's potential profitability at the expense of the factories. He used his association with Rich Fund estate to acquire the Uba de Rich Fund canes and he planted them on his own land. When he had raised enough canes, he distributed cuttings to local small planters. Many of the small planters near Rich Fund, who lived in a cluster of villages named Belvedère, Bon Accueil, Brisée Verdière, Lalmatie, and St. Julien took up the cultivation of Uba, and in 1937 it covered approximately 1,600 hectares there.[58] The Department of Agriculture's cane variety census of 1933 indicated that Uba varieties were spreading among small planters around the island as well.[59] These included several more canes that looked like Uba de Rich Fund; they were either mutations or new hybrid seedlings.

The estates began to worry about the spread of Uba canes. One estate manager recalled in his memoirs that he felt "Uba Marot should be treated as a poison by the factory."[60] The problem was particularly acute among Rich Fund's small planters, but Rich Fund had closed its factory several years before, forcing the Uba growers to bring their canes to three neighboring estates with factories: Sans Souci, Constance, and Union-Flacq. During the early 1930s, the managers of these factories accepted Uba canes, even though they professed to be processing them at a loss. They did so because they hoped to retain the small planters' canes in the event that a rise in sugar prices made Uba profitable to millers. This seemed increasingly unlikely as the depression progressed. Contracts between the estates and the small planters allowed the factories to reject any cane varieties they disliked. The manager of Sans Souci claimed that in 1935 he warned some small planters not to grow Uba canes.[61]

The small planters persisted in cultivating Uba. On July 19th, 1937, small planters brought the year's first harvested canes to Sans Souci's Rich Fund weighbridge. They were shocked to see a sign notifying them that the factory intended to pay 15 percent less for Uba canes than for other varieties: "Planters are warned by these presents that 'Uba' canes generally whatsoever that will be sent to this weighbridge will be subject to a reduction of 15%."[62] The small planters learned shortly thereafter that the factories at Union-Flacq and Constance, the only other nearby factories accepting small planters' canes, had adopted the same policy. Given the depressed state of the sugar market, these Uba growers had been barely breaking even with the Uba canes. Now they would have difficulty repaying their creditors and would have to borrow even more money at higher rates of interest to replant their fields with acceptable varieties.

The small planters began to protest on July 30, 1937. Raoul d'Emmerez, the manager of the Rich Fund estate in Flacq, reported to the Assistant Police Superintendant, R.A. Lavictoire, that his estate's 250 workers, many of whom were also small planters, had walked off the job "for no reason whatsoever." In the midst of an economic depression, labor unrest boded poorly for the production of sugar, the lifeblood of Mauritius. One year later, an official committee of enquiry recorded that Raoul d'Emmerez, like other estate managers, regarded work stoppages not only as a breach of contract, but also as a threat to public order.[63] A strike was more than an economic gesture, it was an attack against the legitimacy of the sugar industry and the colonial state.

Initially the small planters' walkout from their work at Rich Fund estate accomplished very little. Their boss, Raoul d'Emmerez, made it clear to them that he did not sympathize with their cause. Immediately after the 15 percent cut, the small planters asked him to intervene with the factory manager to no avail. Now d'Emmerez and the police reminded the workers that they were not allowed to bargain collectively. Government officials visited the area but did nothing. Two days later the workers returned to the fields, but none brought their canes to the weighing station. Their discontent simmered.[64]

During late July and early August of 1937, a number of labor disputes broke out. On the Chebel sugar estate in the western part of the island, cane cutters left their jobs because of a wage dispute. The Labour Party's leaders, Dr. Maurice Curé and Emmanuel Anquetil, held a rally in Saint Pierre, just west of Rich Fund. On the Constance-Gaieté sugar estate near the east coast, a band of workers interfered with the passage of an estate transport vehicle. Around the same time, major labor disturbances rocked the British sugar colonies of Barbados and Trinidad. Britain's colonial government in Mauritius braced itself for action. Assistant Superintendant Lavictoire of the colonial police visited the Rich Fund area with his supervisor, and listened again to the "exasperated" complaints of the small planters.[65]

Dissatisfied with the police, the small planters took matters into their own hands. After midnight on August 9th, Lavictoire observed a procession of nearly 800 of them marching from his district in the direction of the capital, Port Louis, which lay twenty kilometers to the west. He reported this to police headquarters. The Commissioner of Police, Colonel Deane, drove along the road and intercepted the marchers just outside of Port Louis. Deane invited some of the small planters to meet with him and some government officials at a nearby police station. Once there, a man named Naigum Pallaindya Molain came forward, stating that he was the spokesman of the Labour Party. He handed Deane an unsigned and undated petition, which read:

We, the landowners residing at Bon Accueil, Brisée Verdière and Lalmatie, most respectfully approach Your Honour to inform you that we have during the course of the year borrowed money from our creditors for our business and now as there is a reduction of 15 per cent. on our sugar cane called "Uba," we are unable to make both ends meet. We are by every means reduced to nothing. We have to pay shop-accounts which we find very difficult to defray. We are also very badly paid by the mill-owners. We are thus obliged to live from hand to mouth. We therefore, Sir, most humbly and most respectfully pray Your Honour to find a means for relieving us by some way possible for the betterment in our livelihood.[66]

Another marcher presented the officials with a copy of another petition, with 330 signatures, which had already been sent to the government. It stated:

We, the undersigned planters residing in the district of Flacq in the place called Brisée Verdière - Bon Accueil, beg to lay before you the following for your consideration:
1) We are subject to great injustice by the proprietors of the mill-owners of Sans Souci Sugar Estate, Union-Flacq Sugar Estate, the only two buyers of our sugar canes.
2) The proprietors of these abovenamed estates have created a monopoly this year in proclaiming that our canes "Uba" and canes No.92 will be purchased upon a reduction of 15% because of the lack of sufficient sugar contained in them.
3) The cane "Uba" which the mill-owners have been crushing for the past seventeen years is to-day found to be lacking in sugar, as far as we understand.
4) The attitude of the mill-owners is due:
    1) To the productive nature of the "Uba" in our volcanic land 15 to 20 tons per acre;
    2) To discourage small planters to plant "Uba";
    3) To the public declaration made by the Director of Agriculture to allow mill-owners to give 2/3 sugar extracted to planters.
5) We, poor planters, are relying on the few acres of canes we possess, and hardly we happen to make both ends meet. The reduction of 15% on our canes is really a great loss for us.[67]

The small planters argued that the price cuts were hurting them badly. They called attention to the seemingly arbitrary dislike of the estates' factories for the Uba canes: one year the factory managers accepted them, only to reject them the next. The petitions clearly indicate that the small planters were upset with the high-handed actions of the sugar-milling estates, especially given the depressed state of the economy. The cut in prices was a cruel blow.

At the time, it was not difficult for people in Mauritius to appreciate the broader significance of the small planters' protest, even if some did not sympathize with it. Official records show that government officials recognized that there was more to this protest than feelings of dependency and desperation. Something else was happening, which was showing the close inter-relationship between the colonial state, the Mauritian economy, and the selection of sugar cane varieties. Deane informed the small planters protesting the cut in Uba prices that they constituted an "unauthorised meeting." Collective action was illegal; the small planters were only supposed to make complaints individually to the office of the Protector of Immigrants. Deane was acknowledging that the small planters' protests about Uba canes were more than demonstrations against low prices: the small planters were threatening the colonial order.

The protesters disregarded Deane's advice and continued their march to Port Louis. It was only when Deane confronted them with an armed detachment of police that they accepted the offer of a government train to return them home.[68] But despite Deane's threats, for a time it seemed that the Uba dispute would be solved by negotiation rather than confrontation. Deane and representatives of the Protector of Immigrants met with the protesters in the village of Brisée Verdière, near Rich Fund. The managers of the Sans Souci, Rich Fund, and Constance sugar estates attended this meeting as well. The managers of Union-Flacq, the largest local factory, were conspicuously absent. The small planters reiterated their grievances about the Uba reductions and their labor problems. The manager of Sans Souci estate, Pierre Robert, replied that the planters had been warned since 1935 not to plant Uba because the factory could not make a profit on it. In his opinion, the planters had "utterly disregarded" the warnings.[69]

This unusual dialog between the representatives of the sugar industry, the small planters, and the colonial government resolved nothing. Small planters and workers turned from negotiation back to confrontation, making the estates and the government the targets of their protests. Workers set fires in the fields at Union-Flacq. More groups of small planters and workers attempted to march on Port Louis from different parts of the island. In Saint Pierre, a group of workers menaced the yard of Mon Désert factory. In the village of Saint Julien d'Hotman, near Rich Fund, protesters tipped over four trucks of canes. At the nearby Bel Etang factory, a crowd of six hundred turned over a dozen trucks. In these and other cases, the arrival of the police was enough to end the demonstrations.[70]

The managers of the sugar factory-estates became concerned for their safety. Some suggested that the governor should mobilize the colonial militia, and perhaps even ask the Royal Navy to dispatch a cruiser. These requests show that some Mauritians perceived the small planters' protest to be a serious challenge to the colonial order, but the government decided not to take such extreme measures. The proprietors of the Union-Flacq factory, the Gujadhurs, were particularly concerned for their safety. They were virtually the only Indo-Mauritians who owned a sugar factory and estate, but it is not clear whether this only made them feel more vulnerable. When Ackbar Gujadhur, the estate secretary, asked Police Commissioner Deane what he should do in the event of men marching on his estate, Deane replied, "My only advice is to stop your mill, to mobilize yourselves, and to inform the Police immediately. We will come to protect you." The Gujadhurs construed the word "mobilize" as advice to arm themselves, which they did. They instructed their managers and factory employees to "be ready with rifles and guns," and "to open fire in the event of the mob entering the yard."[71]

On the morning of August 13th, the tense situation in the vicinity of Rich Fund culminated in open violence. Angry groups of demonstrators gathered in several different places even though the police patrolled the area in force. At ten o'clock, Assistant Superintendant Lavictoire found himself at Union-Flacq investigating rumors of an impending attack. Suddenly he saw a group of two hundred small planters and workers on a nearby road, approaching the Union-Flacq factory. He followed the demonstrators into the factory yard, urging them to stop. Near the factory, he saw the Gujadhurs and thirty or forty of their employees, rifles at the ready. An estate overseer approached the crowd in an attempt to speak with them, but instead the crowd pelted him with stones and pieces of cane. Lavictoire succeeded in persuading the crowd to calm themselves. Only then did they begin to select a group of delegates to parlay with the estate staff. At that moment, another group of two hundred planters and workers appeared suddenly on the other side of the factory. They charged the estate staff, throwing rocks, swinging clubs, and yelling "mare salah," which means "beat them" in Bhojpuri. The estate staff withdrew toward the factory, firing their weapons. The crowds ran away, setting fire to the fields as they retreated. When the smoke cleared, four people lay dead and ten people were wounded.[72]

This was the worst violence in Mauritian history. It was also the most serious challenge that working-class Mauritians had ever made against the colonial order. It began as a dispute over prices with the sugar factories, a dispute that was linked inextricably with the production and distribution of a hardy, hybrid sugar cane called Uba. This seemingly contingent dispute spilled over the apparent boundaries of plant selection and distribution, so that it raised questions about the sugar industry and the colonial state. The 1937 riots would force the state to reconstitute its authority, and it did so partly by the distribution of Aimé de Sornay's wonder cane, the SRS's M134/32.

Canes and Order

Historians of the labor movement have portrayed the 1937 riots as an important step in the march toward political power for the masses, particularly those of Indian origin.[73] On the surface, the government made concessions to laborers and small planters, but it really aimed to restore order and to control the strikers.

In 1938, a new governor and a commission of enquiry decided that conciliation, rather than repression, would prevent widespread strikes from ever happening again. This response seemed accomodating on the surface, reflecting the global trends toward anti-colonial agitation and doubt in the old imperialist "mission." On a local level, the Mauritian government began to incorporate non-elite groups within the structures of the state, to guarantee the peace and to preserve a fundamentally unchanged colonial economy. In 1938, the colonial government repealed its ban on labor unions, provided a framework for collective bargaining, and created a Department of Labor to replace the Protector of Immigrants. It also instituted a Central Board to arbitrate disputes between millers and planters. The labor unions had the effect of stabilizing and controlling the workforce, and in most cases the estates secretly welcomed them. The Central Board guarded against unlawful deductions that millers sometimes made from small planters' canes, but it also used chemical analysis of small planters' canes to demonstrate their inferior sucrose content, thus strengthening the case for lower payments.[74]

The colonial government made agricultural extension services one of the centerpieces of this strategy of social control. As early as ten days after the shootings at Union-Flacq, the Director of Agriculture, Gilbert Bodkin, sent the senior chemist of the SRS, Norman Craig, to analyze small planters' soils in Flacq for the first time. He estimated the area under Uba canes and provided small planters with a hardy new Mauritian variety on an experimental basis. Craig noted somewhat belatedly that this might be the "solution to the problem."[75]

The commission of enquiry into the 1937 strikes blamed the unrest on the Department of Agriculture's failure to reach the small planters, despite the explicit instructions of the 1909 commission. The 1937 commission concluded that,

> The Department of Agriculture is the one directly concerned with questions of planting, but the Department does not appear to have taken any action to find a substitute for Uba until complaints began to come in about the cut of fifteen percent. . . . It is clear, however, that the Department of Agriculture will be largely concerned with any solution of the Uba question. We cannot but regret, however, that this question was not taken in hand at an earlier date. The Royal Commission of 1909 made an important recommendation to the effect that the Department of Agriculture should maintain the closest contact with the estates and small planters. That recommendation has not been translated into action so far as we are aware, and we ourselves make a recommendation to this effect. Had the Department of Agriculture been closely in touch with the estates prior to the unrest, the problem created by the Uba cane might conceivably have been avoided.[76]

Officials like Deane had recognized the threat of the small planters to the colonial order. Now, a colonial government enquiry was partially blaming the Department of Agriculture for causing the unrest. Generally, the department was not maintaining close enough contact with the small planters even though the 1909 commission had mandated it. Specifically, the department had done nothing to help farmers resolve the Uba question. It is hard to know what the commissioners meant when they said the department had not been in close contact with the estates; they had been for more than twenty years. This statement was written in the context of a report on small planter problems, and so it is likely that the commissioners thought the department could have gained intelligence about small planters from the estates.

In any case, the Department of Agriculture acted swiftly to replace the Uba canes in Flacq and in other smaller pockets around the island. Governor Bede Clifford may have ordered the department to do so before the report even appeared, at the same time as he purchased more rifles for the police and implemented other repressive and conciliatory measures.[77] So long as the factories did not test for sucrose content, the Uba canes served ideally from the small planters' perspective. After 1938, the Department of Agriculture's new Central Board published a list of "approved" cane varieties that had acceptable yield and disease resistance. It prohibited the cultivation of all other canes, especially the Uba canes.

Before 1937, the state and the estates had dominated the production of plant technology and agronomical knowledge and had limited its distribution. In 1938, the colonial state began to advocate the use of this same technology as a means of staving off a social and political revolution. The department embarked upon a "Uba Replacement Scheme," which immediately raised the question of which cane could be used as a replacement. Was there such a thing as a "small planters' cane," which gave acceptable yields under adverse circumstances, such as lack of fertilizers and nonexistent economies of scale? Or were the same canes good for both the estates and the small planters?

The first efforts of the department to replace the Uba canes indicate that they believed a small planters' cane might exist. One of the first canes introduced to the Uba areas was M108/30, the progeny of a cross between Uba Marot and POJ2878, the Javan wonder-cane.[78] The estates had already rejected M108/30 because of its high fiber content, just as they had rejected the cane's male and female parents. Its vigorous growth suited it to small planters' fields, but it failed the Central Board's tests, and it became one of the first banned cane varieties.

The department experimented with many varieties in the Uba growers' fields. By 1939 it had determined that M134/32 was suited to replace the Uba canes, as well as all the other cane varieties on the island. The broad utility of M134/32 essentially preempted the question of whether such a thing as a small planters' cane could exist, because it was so much better than all other canes growing in the most common conditions. By 1944, the department had distributed new canes to all the Uba growers through the medium of cooperative credit societies and in consultation with the factories. At the same time, the department intensified instruction in the use of fertilizers and established demonstration plots on small planters' lands.[79] Meanwhile, the Central Board discouraged Uba cultivation by allowing the factories to impose increasingly heavier penalties on Uba canes.[80] During the late 1940s and early 1950s, the department was still uncovering rare instances of small planters cultivating Uba canes.[81] One estate manager even complained that Ubas and other similar varieties that were imported from the Coimbatore breeding station in India had potential uses in marginal lands and should not be banned completely.[82] Nevertheless, by the mid-1950s, the department, under pressure from the estates, had succeeded in laying to rest the idea of a small planters' cane, even though some people still believe in it today.

It was not a coincidence that small planters received wider access to new sugar cane technology during the transition to a new state in Mauritius. As Steven Shapin and Simon Schaffer have shown in their study of science in seventeenth-century England, "Solutions to the problems of knowledge are solutions to the problem of social order."[83] For most of the time between 1853 and 1937, the Royal Botanic Gardens, the Station Agronomique, and the Department of Agriculture identified closely with the objectives of the colonial government and the Chamber of Agriculture. Producing new sugar cane varieties became the key to producing sugar, revenues, and a colonial society organized around sugar production. In 1937, when the social order seemed on the verge of collapse, the Department of Agriculture reached out to enroll the small planters into its project of producing and disseminating sugar cane science and technology.

This was not the first time in history that people in positions of authority used the dissemination of science or technology as a means of deflating social unrest, nor was it the last. But Mauritian history does force a reconsideration of the ways in which historians have addressed the relationship between science, technology, and imperialism. Daniel Headrick has argued that for the period up to 1940, Europeans restricted the distribution of technical knowledge as a means of fostering colonial dependency.[84] But in Mauritius, in the aftermath of the Uba riots, the British colonial government actually distributed technical knowledge, in this case as a means to control people. This does not necessarily contradict Headrick's point: ever since the middle of the nineteenth century, the colonial government and the sugar estates did withold technical knowledge and sugar cane varieties from small planters. But the riots of 1937 and the government's response indicate that there may be other patterns in the history of science, technology, and imperialism.

In Mauritius, local politics and economics influenced natural resource management and agricultural science in significant ways. Colonial state officials and the sugar industry's elite cooperated extensively in the production and distribution of the knowledge and technology of sugar cane cultivation. Before 1937, they had the power to exclude small planters from equal access to this knowledge and technology. As a consequence, sugar cane breeders, the most crucial group of agricultural scientists on the island, failed to establish their own credibility outside of the sugar estates. In 1937, when Mauritius began a thirty-year process of decolonization and democratization, the small planters' struggle for access to knowledge and technology came to the forefront. The state and the elite learned a paradoxical lesson: that they could distribute sugar cane varieties, the lynchpins of the economy, as a way to quell social unrest.

Immediately after 1937, when Mauritian decolonization began, the distribution of sugar canes abetted colonialism. This raises a troubling question in the history of Western efforts to "develop" non-Western countries, many of which are former colonies. During the middle of the twentieth century, the rise of a development ideology among Westerners coincided with decolonization, but were Western efforts to improve the living standards of non-Westerners thinly-disguised methods of "pacification" that served ultimately to prolong colonial economic exploitation?

In 1937, the rise of anticolonialism and the depressed condition of the world economy provided the backdrop to the riots, but the protesters at Rich Fund and Union-Flacq drew attention not only to a reduction in payments, but also to the sugar estates' ban on the cultivation of the Uba cane variety. The 1937 "Uba riots" provide a window through which we can see how people in a colony produced and distributed agricultural knowledge and techniques in many different ways.

The Uba riots are but one reminder that the distribution of sugar cane varieties was an integral part of the society, economy, and polity of both Mauritius and the British Empire. Mauritius and Britain's other sugar-producing colonies depended on the production of new canes for their survival. In turn, the cane plant itself gave structure to their economies and societies. The production and distribution of new canes was not value-neutral, but reflected the often conflicting objectives of the state and non-governmental organizations, as well as the relative influence of different classes of cane-growers. Each sugar cane plant growing on the island represented a set of political and economic choices. It is not surprising, then, that in Mauritius and other sugar islands, political and economic protest often took the form of burning cane fields. This was an attack that struck at the planters' pocketbooks. It was also a gesture against everything the sugar cane plant represented.

6

SUGAR AND THE MAKING OF A NEW STATE, 1937–1993

Given the long-term importance of the production of cane varieties, it is not surprising that a dispute over the plants should have helped to cause the riots that swept through rural Mauritius in 1937. Beginning in the middle of the nineteenth century and continuing through the early twentieth century, the estate-owning elite had joined with the government in creating and maintaining agricultural research institutions as a way to boost productivity. Everyone involved in sugar production contributed funds for research, including small planters, but the owners of the large estates retained a dominant influence over the research agendas of scientific institutions. At the same time, Franco-Mauritians and assimilated Creoles began to occupy some of the most important posts in the Department of Agriculture and the SRS.

Small planters challenged the Chamber and the government's assumption that everyone derived prosperity from estate-supported scientific initiatives. Small planters lacked access to the newspapers and other public forums of debate because of widespread discrimination and illiteracy. However, small planters still worked to undermine the conventional wisdom that the state and the estates exercised in scientific matters. They themselves drew upon multiple sources to produce their own knowledge about sugar cane plants. The small planters' resistance manifested itself most clearly during the riots of 1937, and afterwards.

The Collapse of the SRS

The 1937 riots occurred at a time when Britain's government was beginning to entertain serious doubts about its colonial empire. Riots broke out in the West Indies during the same year, while Hitler was using the existence of the British and French empires as an embarrassing excuse to promote German *lebensraum* in Eastern Europe.[1] The new governor of Mauritius, Bede Clifford, took cautious action upon the recommendations of a commission of enquiry. In 1938, he created a new labor department, appointing six inspectors to mediate between workers and the estates. He also named two Indo-Mauritian small planters to the Council of Government, and the council passed legislation guaranteeing minimum standards of health and nutrition for workers and payment of wages in cash. The Central Board would now mediate disputes between small planters and millers. New legislation also permitted workers to form unions, although island-wide unions and syndicates were not allowed. Despite government concessions, the situation remained tense. Labour Party union organizers complained about this restrictive labor legislation, but the government had little patience with Curé and his associates. By 1940, the Labour Party virtually ceased to exist because of government harassment.[2]

The Second World War reinforced the island's dependency on the British government's willingness to provide a market. When war broke out in Europe in 1939, the British government suspended trading on domestic sugar markets. Then it undertook to purchase and ration all colonial sugar at a fixed price, to ensure availability and price stability. Each year, it increased prices paid to producers, but these fell short of world market rates. This arrangement lasted until 1949, by which time the Mauritius Chamber of Agriculture and representatives of other colonies' sugar industries were clamoring for change.

Given the tensions of the 1930s and 1940s, it cannot be emphasized enough how important it was that the SRS produced its hybrid wonder cane, M134/32. It tolerated *Clemora*, it resisted most diseases, and it yielded well virtually everywhere in Mauritius. Only in the 1940s did it show itself to be somewhat susceptible to red rot and the new viral disease called chlorotic streak. The SRS released M134/32 in 1937 and by 1945, the hybrid accounted for 37 percent of the area under cane. In 1951, it reached its peak of popularity, with 91 percent.[3] It increased the entire island's yield of canes per hectare from a five-year running average of 49.7 in 1937, to 60.8 in 1951.[4] M134/32 raised the island-wide percentage of commercial sugar recovered from cane, from a five-year running average of 11.4 in 1937, to 11.6 in 1951. During the same time, average laboratory sucrose extraction rose moderately too, from a five-year running average of 13.3 percent in 1937, to 13.5 in 1951. (See Table 4.) These figures meant that cane growers could obtain more sugar per hectare from M134/32; the increase of cane tonnage in the field did not merely represent an increase in fiber or bagasse. Given the labor unrest and the economic uncertainties of decolonization, increased productivity could not have come at a better time for the estates.

Knowing that it was dangerous for the sugar industry to depend on one variety, the breeders at the SRS did not rest on their laurels after the original release of M134/32. Geneticists continued to breed hybrid canes and they also introduced more extensive methods. They could still obtain 75,000 seedlings from 150 crosses, but they extended the trials at Réduit from four to six years, before selecting about four of the new seedling varieties to be sent to the estates for trials.[5] They added the important attribute of cyclone resistance to the established criteria of high yield and resistance to pests and diseases.[6]

Even the Second World War did not interrupt the breeding program in any significant way. Most of the staff of the SRS and the Department of Agriculture were seconded for war duties of various kinds, but the colonial government allowed full-time breeding work to continue, recognizing its great importance.[7] Nevertheless, none of the canes bred at the SRS after M134/32 attained any significant popularity before 1953, when the government and the sugar industry allowed the new Mauritius Sugar Industry Research Institute (MSIRI) to absorb the SRS. This does not mean that the SRS research was unproductive: between 1955 and 1974, the MSIRI released many SRS-bred varieties, with ten attaining moderate popularity.[8]

Despite its successes in cane breeding and its continued usefulness in agronomy, the SRS faced significant institutional problems. The depression limited the SRS budget, but hard times in Britain allowed the colonial departments of agriculture to recruit some of the best university graduates.[9] During the war, recruitment of agricultural officers came to a halt. Those officers who were present in Mauritius took on additional responsibilities, such as Chief Censor and Malaria Control Officer, duties that limited the time they could devote to agricultural affairs. During the postwar economic boom, many of the senior officers resigned their positions when they realized that they could earn significantly greater salaries in the private sector. As early as 1947, the Director of Agriculture complained about curtailing sugar cane research because of insufficient staff.[10] By the early 1950s,

Table 4

Sucrose Content of Mauritian Sugar Cane, 1811-1953

Year	Average Percentage of Sucrose in Cane (Factory Extraction)	Year	Average Percentage of Sucrose in Cane (Factory Extraction)	Average Percentage of Sucrose in Cane (Laboratory Extraction)
1811-1820	5.00	1927	10.53	—
1821-1830	6.00	1928	10.76	—
1831-1840	6.75	1929	10.89	13.08
1841-1850	7.10	1930	10.92	12.57
1851-1860	7.60	1931	10.61	12.97
1861-1870	8.10	1932	11.09	12.73
1871-1880	8.20	1933	11.25	13.24
1881-1890	8.30	1934	11.02	12.86
1891-1900	8.70	1935	11.21	13.14
1901-1908	10.10	1936	11.79	13.69
1909	10.56	1937	11.12	12.99
1910	10.50	1938	11.94	13.87
1911	10.66	1939	10.81	12.71
1912	10.42	1940	11.45	13.29
1913	10.76	1941	11.49	13.30
1914	10.62	1942	11.81	13.67
1915	10.83	1943	12.11	13.87
1916	10.30	1944	11.31	13.42
1917	10.62	1945	11.14	13.42
1918	10.95	1946	11.50	13.56
1919	10.32	1947	12.41	14.37
1920	10.76	1948	12.36	14.31
1921	9.90	1949	12.40	14.33
1922	10.58	1950	12.27	14.14
1923	10.51	1951	11.12	13.03
1924	10.28	1952	11.42	13.26
1925	10.56	1953	11.03	12.96
1926	9.94			

Sources: The Mauritius Chamber of Agriculture, *The Mauritius Chamber of Agriculture, 1853–1953* (Port Louis: General Printing and Stationery Co., 1953); and the "Tableaux synoptiques," *Revue agricole et sucrière de l'île Maurice*, 1928–1953.

the SRS lacked a senior botanist and a senior chemist, and had little hope of acquiring replacements given the low government pay scale. The breeding program received top priority at the expense of other programs.[11]

The senior SRS staff also misunderstood the Mauritian estate managers and owners, compounding their institution's problems. Senior staff members were all British, and with a few exceptions, they took the colonial position of distancing themselves from Mauritians. Many never even learned to speak French or Kreol, and most SRS publications were in English, in spite of the fact that the majority of estate staff members preferred to read French. The SRS staff seemed aloof, even though they relied heavily upon the estates for funding and for access to trial plots. For example, in 1941 the chief cane breeder, G.C. Stevenson, gave a speech to the alumni of the College of Agriculture, many of whom worked in the sugar industry. He remarked that the interests of the SRS staff were "academic, rather than financial," but coolly assured his audience that it was "a source of encouragement to us to see the way in which the results of our work are put to practical use."[12]

Stevenson was making a worldwide reputation for himself as an expert on sugar cane genetics, but locally he had trouble with the sugar industry. Under his guidance, the SRS continued to use Uba Marot as a parent cane in its breeding program. In 1947, Stevenson proposed releasing three more of Uba Marot's progeny, but he ran into opposition. Pierre Halais, who had formerly worked as an assistant breeder under Aimé de Sornay as well as working as the SRS's junior chemist, passed along a tip to the Chamber: these three canes with Uba Marot in their ancestry had a relatively low percentage of sucrose in the extracted juice. The Chamber complained loudly. Stevenson retorted that it was dangerous for Mauritius to depend on one cane, even M134/32, and that some plantings should still be made of the Uba's descendants. The government sided with the Chamber and prohibited the distribution of these canes.[13]

Previous sugar cane research institutions like the Pamplemousses gardens and the Station Agronomique foundered in part because relations with the sugar estates and the Chamber became rocky. The estate-owners held great political clout in colonial Mauritius, but sugar cane researchers also relied on them as witnesses in establishing their own credibility. The failure of SRS researchers to cultivate a close working relationship with the Chamber, together with the rise to power of non-elite Mauritians, undermined the institution.

As early as 1933, some prominent members of the Chamber began to criticize the SRS. According to a confidential source, Pierre de Sornay and five estate managers brought grievances to the Chamber's leadership alleging sloppy research work at the SRS. Not only were some experiments designed poorly, they said, but the SRS staff did not take criticism from the estates very well. They also accused the British senior staff of taking credit for work done by the Mauritian junior staff and complained about how the SRS published its results in complex, jargon-filled English rather than in simple French. They suggested that the Chamber create a committee of estate staff members to supervise the work of the SRS, harkening back to the days when members of the Chamber supervised the work of the Pamplemousses gardens. Pierre de Sornay's dispute with the SRS over research results spilled over into the pages of the *Revue Agricole*, with the SRS botanist and chemist arguing persuasively in their own defense. Their articles forced Pierre de Sornay to withdraw his accusations of sloppiness, but the Chamber still endorsed the idea of creating a supervisory committee.[14]

Once again, the Chamber's patience with government agricultural researchers was wearing thin. At the time of Pierre de Sornay's dispute, M134/32 had not yet been released, and some of the most prominent estates were resorting to the illegal and dangerous practice of smuggling new cane varieties from abroad.[15] The Chamber was even reminding the SRS that Tempany had only introduced BH10(12), then becoming popular on the island, because of pressure from an estate proprietor.[16] In 1935, the Chamber persuaded the governor to create an SRS Research Advisory Committee composed of the Director of Agriculture, the head of the SRS, and members of the Chamber.[17] The records of the discussions leading to the creation of this committee are not available, but at that time, it was still common for the Chamber to claim that its members were paying for the SRS and therefore had a right to influence its research, even more than they could through the board of the Sugar Industry Reserve Fund. The period 1937–1943 saw several outbreaks of labor unrest, as well as the beginnings of agricultural trade unionism, and the Chamber was aware that improved cane varieties and better agronomical practices could contain possible rises in the costs of production. The Director of Agriculture, Gilbert Bodkin, may not have fought the Chamber's intrusion onto his own turf. Although he was a noted administrator, Bodkin was not as forceful a personality as Tempany.[18]

The sugar industry sought to control agricultural research outright because it did not feel that the SRS was solicitous enough of its needs. During the war, confidential sources indicate that once again the Chamber heard complaints about the SRS from the sugar estates.[19] In 1945, the Chamber began to consider the possibility of the sugar industry taking over the work of the SRS. Neither the formation of the Research Advisory Committee, nor the release of M134/32 was enough to satisfy them.

The estate owners began to endow their own small independent research institutions because they felt disgruntled with the limited effectiveness of public institutions. In 1946, the Sugar Industry Reserve Fund established an agricultural chemistry laboratory under the direction of Pierre Halais, formerly of the SRS. In many ways he represented the intersection of the communities of the sugar industry and the government agricultural scientists. His purpose was to conduct foliar diagnoses, a task for which the department had always been responsible. In fact, even the SRS began to avail itself of the laboratory's services.[20] In 1947, seven estate owners in the north grouped together to finance a Station Agronomique du Nord and hired a young Franco-Mauritian agronomist named Guy Rouillard to conduct research. He came from a prominent estate-owning family and like Stevenson had received his training in the Imperial College of Tropical Agriculture in Trinidad, where colonial agricultural officers studied. He had credentials from colonial science and Mauritian sugar production. He performed fertilizer and herbicide trials and investigated new cultural practices. He did some of these in conjunction with the SRS, but most were independent efforts.[21]

Some estates even took up cane breeding themselves. As early as 1937, the Anglo-Ceylon Company's Highlands estate produced a useful cane called Ebène 1/37. By 1960 this variety had reached a maximum distribution of 26 percent of the area under cane.[22] Such independent efforts could not support an entire industry, but neither could an SRS that was, in the words of William Allan, the new Director of Agriculture, "severely handicapped by the lack of senior officers."[23] The Chamber, for its part, decried how the "talented" SRS staff received "ridiculous" salaries, considering the importance of their work to the industry.[24] The progressive worsening of institutional problems at the SRS disappointed the Chamber, which in turn hampered the institution even further.

Founding the MSIRI

During the late 1940s, the Chamber began to contemplate new ways to build institutions for sugar cane research, but its members had to take into account a rapidly changing political scene. Anticolonial politicians were gaining more support in Mauritius and more sympathy in London, even as they shifted from a rhetoric of class struggle to communal identity as a basis for anticolonial solidarity. In 1945, the accession of a Labour government to power in Britain caused significant reversals in Mauritian labor policy. The colonial government now provided an official adviser to the

Mauritian unions, which regained their strength within a year. By this time, employers were recognizing that unions provided a useful tool for stabilizing the workforce, even if they made the occasional "unreasonable" demand.

The prime driving force behind this transformation was a movement to revive Hindu culture, under the leadership of Basdeo and Sookdeo Bissoondoyal. Beginning in the early 1940s, the Bissoondoyals taught the Hindi language and Hindu culture in village gathering places called *baitkas.* They formed a large organization, whose influence spread from its original base in the south to the entire Hindu community of Mauritius. They fostered pride in being Hindu, building up the self-respect of many poor laborers and small planters. They also made the Hindus' Muslim and Creole neighbors extremely nervous because of the ongoing communal violence in India. Hindu intellectuals were also playing a more prominent role within the established political arrangements, particularly after 1940, with the appointment of a young, vocal doctor named Seewoosagur Ramgoolam to the Council of Government.[25]

Indo-Mauritian and Creole intellectuals continued to press for constitutional change and found that the government was increasingly willing to listen. One governor, Donald Mackenzie-Kennedy (1942–1949), worked conscientiously for reform, even though the Colonial Office regarded Mauritius as a low priority. Between 1945 and 1947, he floated three separate constitutional proposals after consulting with various committees of Mauritians and encouraging debate in the press. Discussions focused on how to guarantee the rights of minority communities within a truly democratic framework, a strong indication of how the political terrain had shifted from class back to community. The Franco-Mauritians and Creoles, with a combined total of about 30 percent of the population, tended to support proposals that would discount the value of individual Hindu votes, such as literacy requirements and separate electoral rolls to elect representatives for each community. They recognized that Hindus made up about 51 percent of the population, and if unified could dominate any elected assembly. The Muslims accounted for about 16 percent of the population, and their leaders vacillated on questions of communal representation. Hindus such as Ramgoolam and the Bissoondoyals insisted on a strict "one man, one vote" system.[26] Ironically, in 1947 the governor ended the debate by using his absolute powers to impose a new constitution. This document extended the franchise to all Mauritian men and women over 21 capable of writing a simple sentence in any language. The new Legislative Council was made up of nineteen elected and fifteen nominated members, three of whom would be government officials, while the governor retained veto powers.[27]

The new constitution was a clear victory for the democrats and Hindus. The 1948 elections to the Legislative Council demonstrated this quite clearly. Hindu professionals, including a group of six centered around Ramgoolam, won eleven of nineteen elected seats, while Creoles won seven, and a Franco-Mauritian liberal won one. Ramgoolam formed an alliance with Labour against the conservative block appointed by the governor, and by the time of the next elections in 1953 Ramgoolam emerged as one of the recognized leaders of the Labour Party. The governor allowed elected members to sit on the Executive Council, the equivalent of a cabinet. In 1950, the governor made each unofficial member of the Executive Council a legislative "liaison officer" with a government department, the first step toward ministerial government. Members became fully responsible for their departments in 1957.[28]

As the Labour Party grew more powerful, small planters gained representation in the debates about creating agricultural research institutions. General dissatisfaction with the SRS led to renewed cooperation between the Chamber and the government in creating an institution to provide the sugar industry with plants and information. Representatives of the Chamber met with the governor and sent letters to the government urging the improvement of salaries at the SRS. In 1947 and 1948, when the Colonial Office sent a commission to investigate the Mauritian economy, the Chamber and the government insisted that it address the question of improving agricultural research. The commission agreed to look into the SRS, and its approach reflected the contemporary policy of gradual, negotiated democratization. It formed a special panel called the "Efficiency of Sugar Cane Production in Mauritius" committee made up of representatives of the Chamber, the Department of Agriculture, the small planters, and the trade unions. This was the first opportunity for the small planters to speak out in an official discussion of agricultural research. The committee agreed that the existing research situation at the SRS was unsatisfactory and debated whether to reorganize and ameliorate the SRS, to abolish the SRS but institute a new research branch within the Department of Agriculture, or to transfer the SRS to the sugar industry.

The committee's discussions proved tentative at first, but continued problems at the SRS moved the participants to take more radical steps. At first the Chamber hoped to improve upon the SRS because a complete industry takeover of research was likely to prove costly and complicated. But in early 1948, two of the best-liked senior staff members of the SRS announced they were transferring to posts in other colonies. The Chamber decided that this was the last straw. Its representatives worked out an agreement with the government to the effect that research would flourish best under the industry's aegis. But the small planters' representatives feared discrimination at the hands of the Chamber's scientific appointees. The small planters continued to favor improving the SRS, but it was unclear where the funding would come from to make the necessary changes. The small planters were outvoted on the committee, but still managed to obtain some concessions. The industry's new research station would have to recruit its staff impartially, and to make certain of this, representatives of the small planters would sit on the institution's board of directors. The Department of Agriculture also agreed to consider expanding its small planter extension services.[29]

Despite the problems of soliciting state support for agricultural research, the estates hoped to benefit from continued state involvement, particularly in finance. In 1948, the Chamber gained virtual control over research, but reversed its position and balked at the cost of expanding the SRS. The Chamber's president made excuses, explaining that certain improvements in the SRS had already taken place: the government had just constructed a new building for the SRS, so it could move out of its quarters in the basement of the College of Agriculture. In 1949, the Colonial Office also initiated its Colonial Research Service, and the Chamber hoped some of its new officers would serve in Mauritius. The Chamber approved of the new cane varieties being distributed. It also considered Allan, the new Director of Agriculture, to be a very "competent, energetic, and zealous" man who was likely to improve matters.[30] For his own part, Allan tried to persuade the Chamber that they ought to take over the SRS. When they refused, he began entertaining proposals to amalgamate the SRS with other research institutions in East Africa and even French Réunion, but none of these would benefit the peculiar crop research needs of Mauritius. The United States Economic Cooperation Association even suggested a loan of some American scientists to the SRS, but this plan never came to fruition. In the meantime, the SRS continued to decline.

The government persisted in pressuring the sugar industry to take over the SRS. In 1950, the governor joined Allan in calling on the Chamber to reconsider its fear of financing a research institution. In a speech dedicating a new building at the SRS, the governor recited the litany of problems there and argued that if the industry assumed responsibility for research, not only would it obtain better results, like other colonies with private research stations, but the government could concentrate its efforts on other agricultural projects.[31] On the surface, it seemed that the government stood to lose some authority in divesting itself of a branch of the

Department of Agriculture. In reality, creating a non-governmental research organization was a way of subtly expanding the state's authority. On the one hand, the government could use the funds and staff it usually allocated to the SRS for other projects. On the other hand, a more efficient research station was likely to increase production, and therefore ensure more revenues for the expanding welfare state. Government representatives would sit on the new organization's board. In any case, the government could make life uncomfortable for any potentially rebellious scientists or industrialists, because it levied and collected the cess on sugar exports that funded research.

Thanks to the government's efforts, the Chamber once again saw potential advantages in a private research station and changed its position to favor an institution run by the industry. But it proceeded very carefully. In 1950, it dispatched Pierre Halais, the chemist in the foliar diagnosis laboratory, to visit the private sugar cane research stations in South Africa, Australia, and Hawaii. Little is known about how the Chamber reached its decision, but in his annual report for 1950, the president emphasized how local problems were forcing a reconsideration, although inquiries in foreign countries seem to have reassured him that a private research station could work.[32]

In early 1951, the Chamber formally notified the governor that it sought the transfer of the SRS to the sugar industry, but democratization threatened the traditional process of negotiation between the Chamber and the government in scientific matters. The local political situation had changed a great deal since the Chamber and the government created the Station Agronomique in 1892. The 1947 constitution had created a Legislative Council with nineteen democratically elected members and fifteen appointed members. Any proposal to transfer the SRS would have to receive the Legislative Council's approval, and during the 1947–1948 Economic Commission, many of the elected members, including Ramgoolam, the leader of the Labour Party, had indicated their opposition. In 1951 and 1952, while a committee of representatives of the Chamber and the government discussed the details of the transfer, a showdown was brewing in the Legislative Council. The Chamber felt fairly confident it had enough votes to achieve the transfer if it could obtain passage of a bill before the elections of 1953.

By the middle of 1952, the committee on the transfer of the SRS worked out a plan to create the Mauritius Sugar Industry Research Institute (MSIRI), which followed the recommendations of the 1947–1948 commission. The government would levy an export tax on sugar to support the MSIRI, much as it had done for the Station Agronomique, and it would sell the SRS buildings to the MSIRI. An administrative council would supervise the new institute, and its members would come from the government and the Chamber, while including representatives of estates with and without factories, and small planters. The committee secured the broad approval of all the island's major sugar-producing organizations, including the Mauritius Planters' Association and the Federation of Cooperative Societies, which represented small planters.

A consensus emerged among the members of the committee that Paul Octave Wiehé, a Franco-Mauritian, would be the best person to serve as the MSIRI's first director. Wiehé had the credentials and the experience to bridge the gap between colonial scientists and the sugar industry. Wiehé graduated first in his class from the Mauritius College of Agriculture in 1930, and accepted a scholarship to study at the Imperial College of Science and Technology in London, where he earned a B.S. in botany in 1934. After graduation, he studied pasture grasses at the Aberystwyth Plant Breeding Station, then returned to Mauritius to teach biology at the Royal College and the College of Agriculture. In 1938, he joined the Department of Agriculture, where he worked as a phytopathologist and made numerous contributions to Mauritian botany. Most Mauritians who worked for the department stayed in Mauritius, but in 1948 Wiehé decided to take a job as the colonial plant pathologist in Nyasaland.[33]

The process of recruiting a director for the MSIRI caused the Chamber to increase its commitment to the existence of a private institute. Wiehé agreed to direct the MSIRI on one major condition: if he was going to resign from the Colonial Agricultural Service to become director of an institution that did not yet exist, he wanted some assurance that the MSIRI would indeed be created. The Chamber contacted the estate owners, who all agreed that if the MSIRI Bill failed in the Legislative Council, they would subscribe funds to create their own research institution. This would be difficult and expensive to do without government support, but either the Chamber felt Wiehé was worth the trouble, or it was gambling that the MSIRI bill would pass.

In early 1953 Wiehé, the Chamber, the colonial secretary, and delegates from other producers' organizations discussed the final details of the MSIRI bill. The Chamber negotiated from a position of strength because the estates were willing to support their own institute. The specific composition of the MSIRI's administrative council became a bone of contention, but in the end the Chamber achieved strong representation on it. As proposed to the Legislative Council, the council would include one representative from the Chamber, one from the government, three from the estates with factories, one from the estates without factories, and two from the small planters.

In 1953, the Legislative Council passed the MSIRI Bill, over the vehement opposition of the Labour Party. Ramgoolam thundered that "I have not yet enough confidence in the rectitude and correctness of the capitalists who run the sugar industry from what we have seen from experience in the past. . . . They run it for their benefit, not for the benefit of the [small] planters. . . . they do not seem to run things nationally as is their pretention." The Labourites accused D. Luckeenarain and R. Balgobin, the members of the Legislative Council who also represented the small planters' Mauritius Planters' Association in the MSIRI negotiations, of allowing the estate owners to cow them into accepting a package that was not in the interest of the small planters. Sugar cane researchers were now reaping the bitter harvest of neglecting to establish their credibility with small planters. During the debate, a member who was the former president of the Chamber defended the MSIRI Bill, arguing that it was impossible to create a new cane variety that would not benefit all sectors of the sugar industry. When the debate ended and the members voted, Luckeenarain and Balgobin maintained their support for the MSIRI. After a vote of fourteen to ten, the new institute could begin its work.[34]

The alliance between the Chamber and the government had once again created a new research institution to serve the sugar industry. Had they waited any longer, the processes of decolonization and democratization might have stymied its founding. Ramgoolam's concerns about the dissemination of science and technology to small planters had deep historical roots. Research institutions, created through bargaining between the estate-owning elite and the colonial government, had usually ignored the small planters. Already in 1953, the Chamber and the government needed the support of the small planters' organizations to pass the MSIRI Bill. This was a direct outcome of events in 1937, when small planters contested the colonial order as part of a protest about the Uba cane variety.

Decolonization and Corporatism

Young professionals educated in England were taking charge of the Mauritian anticolonial movement. They drew much of their inspiration from Fabian social democracy, and during the 1953 campaign the revitalized Labour Party adopted a typical socialist platform. Its members sought to nationalize the sugar industry, to build more roads, schools, and hospitals, and to implement national health insurance and food subsidies. But the party also came to be identified with the interests of the Hindu community, and a somewhat anti-Hindu opposition of Creoles, Muslims, and Franco-Mauritians coalesced and called themselves the Ralliement Mauricien. The Labour Party won thirteen of the nineteen elected seats, and the Ralliement only two, but once again the governor appointed enough conservatives to the nominated seats to offset the Labour victory.[35]

The governor's cancelling-out of the Labour majority caused great consternation, but during the late 1950s Britain committed itself more firmly to granting independence to its colonies. A long series of conferences and committee meetings took place in both London and Mauritius, in which many of the same issues of communal versus direct representation were raised. Over the course of these discussions, the Labour Party convinced the governor and the Colonial Office that the majority of Mauritians would not accept anything short of a one-man, one-vote system. The resulting constitution of 1958 removed the literacy requirement for the vote and ended the system of appointing some members of the Legislative Council. Starting in 1959, voters could elect all of the Legislative Council, while the governor retained the power to appoint the most popular "best losers" to the legislature, in order to ensure the representation of all communities.[36] The principles set forth in the 1958 constitution remain substantially unchanged today, although in 1967, the National Assembly expanded from 40 to 70 seats.

During the 1960s, communal politics intensified. In both 1965 and 1968, communal violence flared up, and British troops landed to keep the peace. But the British government was determined to grant independence. By the mid-1960s, Britain had already divested itself of most of its colonial possessions, and saw no reason to delay doing so in Mauritius. In 1961 and 1965, negotiations took place between representatives of the Colonial Office and the Mauritian political parties to resolve the specific terms of independence. Mauritius became independent on March 12, 1968, with Ramgoolam as the first prime minister, but with the Queen of England continuing as head of state, represented by a governor-general.

Most accounts of Mauritian history argue that between 1938 and 1968, the island's government grew more democratic and representative. This was true on the surface. Under the constitution of 1886–1947 two percent of the population elected a minority of the Council of Government, while after 1968 the entire adult population could elect a parliament independent of outside control. Mauritians also take pride in the fact that they achieved these changes with a minimum of violence, especially when compared with the experiences of other colonies.

Nevertheless, judging from the perspective of 1968 the historical continuities are more striking than the changes. Sugar cane culture and the Franco-Mauritian elite still dominated the island's economy. The sugar industry remained divided as before between large-scale and small-scale farmers. In 1968, it accounted for 93 percent of exports, 94 percent of cultivated land on 47 percent of the island's land surface, and it employed 30 percent of the island's work force.[37] The revenues from sugar exports continued to prop up all government expenditures. The newly independent government could accomplish very little without a healthy sugar industry.

Decolonization and the creation of a welfare state were changing many aspects of sugar production. As centralization progressed, the symbiotic system of small planters and large estates continued in place. But democratization sapped the political power of the industrial elite and increased the amount of legal restrictions placed upon the industry. After 1937, the government mediated disputes between millers and planters and permitted limited forms of labor unions. Unions gained major concessions from the industry as the Labour Party's power grew. In 1946, the government also compelled the industry to subscribe to complex crop insurance and pension fund schemes. Between 1937 and 1988, the government required factories to increase payments to small planters. Taxes on sugar exports also rose, and by the 1970s they were falling disproportionately heavily on the estates.[38]

Decolonization may have gradually shut out the parliamentary participation of the sugar industry's delegates, but it did not diminish the industry's ability to extract major concessions from the governments that regulated it. The most important of these involved marketing sugar. In 1949, the British government assumed that the world sugar shortage would end in several years, and announced that it would continue to buy all colonial sugar until 1952. Then it entered into negotiations with the colonies for future sugar purchases. Britain had two reasons to seek social stability in the sugar colonies. A decent price for sugar would ensure a steady supply, but it would also have the effect of reducing the immigration of colonial subjects to Britain. In 1951, after two years of wrangling, Britain and the colonies signed the Commonwealth Sugar Agreement, which provided each colony with a quota of sugar that could be sold on the British market at a remunerative price to be negotiated annually on the basis of production costs.[39]

The agreement proved beneficial enough to Britain and the colonies that it continued in force until 1974. The negotiated price and the quotas for Mauritian sugar provided a stable market for most of the island's production. It reduced the island's geographic vulnerability considerably, while allowing for more long-term economic planning. Neither the Commonwealth Sugar Agreement of 1951, nor the successor Sugar Protocol of the Lomé Convention in 1975, could have been negotiated so successfully without close cooperation between the sugar industry and the government. The industry needed a high price in order to renew itself and expand, while the government needed a healthy and taxable sugar industry. Both needed domestic tranquility. These agreements ensured the continuing profitability of the sugar industry, even if the price of prosperity was reliance upon negotiations with foreign governments to set the price of sugar.

The sugar industry adapted to major shifts in Mauritian, British, and world politics. During decolonization, the British colonial authorities and their Mauritian adversaries added a representative government onto the surface of what had always been a "corporatist" government. In other words, this was a government that reached the most important decisions outside of representative institutions, working in tandem with nongovernmental economic organizations such as the Chamber of Commerce, the Chamber of Agriculture, and later, the trade unions. This mirrored the way that western European governments reconstructed their authority after both world wars. The corporatist approach was designed to shore up bourgeois values, but in doing so sacrificed the liberal bourgeois principle of a representative government that did not interfere in the private sector.[40]

The shift to an independent government added a new factor to the corporatist equation. The Franco-Mauritian elite vanished from electoral politics, while continuing to concentrate on agriculture and commerce. In addition to their traditional economic role, during the 1970s they provided the driving capitalist force behind the new tourist and textile industries. The Chamber of Commerce and the Chamber of Agriculture became more diverse, but they felt threatened by the socialist rhetoric of the majority political parties. The success of capitalism in independent Mauritius vindicated their position. It is a poorly kept secret that many socialist politicians seek the advice of the two Chambers when making the most important economic decisions.

The politics of independent Mauritius have been a complex mix of alliances based on community and class, tending toward the peaceful resolution of disputes in a democratic manner. Between 1968 and 1982, Ramgoolam and the Labour Party held the upper hand in coalition governments. Labour came to be known as a Hindu party, and used its political influence to staff much of the civil service with its followers. Muslims and Creoles tended to identify with other parties, while Franco-Mauritians and

Sino-Mauritians remained aloof. Despite communal political allegiances, all the parties shared a commitment to Fabian social democracy and implemented welfare-state policies. The most vocal opposition came from the Mouvement Militant Mauricien (MMM), which combined radical socialist ideology with trade unionism in an effort to forge a class-based, cross-communal party.[41]

The early years of independence were a difficult time for the island, testing Mauritians' commitment to democracy. In 1972, the MMM's revolutionary threats and involvement in strikes persuaded Ramgoolam to declare a state of emergency. Lasting until 1976, this measure suspended elections and repressed all political activities. Sugar prices were high during the emergency, and the government and the industry took advantage of the situation to diversify the economy into tourism and textile manufacturing. Prosperity led the MMM to moderate its rhetoric somewhat and to reaffirm its commitment to democratic politics. Its nationalist ideas captured the imagination of people from all communities, particularly young people. The MMM won a plurality in the 1976 elections, although a Labour-based coalition in the National Assembly kept it out of power. However, during the late 1970s the economy soured and the Labour Party endured several corruption scandals. Overpopulation ensured high unemployment. To make matters worse, the International Monetary Fund (IMF) forced a painful program of structural adjustment on the country. In the elections of 1982, popular resentment against Labour and the IMF swept the MMM into power.[42]

Despite the rise of the radical MMM in popularity, the 1980s and early 1990s saw all political parties adopt a procapitalist stance. In 1982, it only took a few weeks for the IMF to persuade the new MMM government of the value of economic austerity. Since then, coalition governments have come and gone. The Labour Party and the MMM remain the most important coalition partners, but they have also formed coalitions with the Mouvement Socialiste Mauricien (MSM) and the Parti Mauricien Social Démocrate (PMSD) as well as with smaller parties.[43] The island's continued economic success helped to ensure that few major changes would occur. Today one can attend May Day rallies for these parties, self-described "militants" and "socialists," and listen to their political leaders taking credit for the success of Mauritian capitalism. One of the most important changes was largely symbolic: in 1992 Mauritius became a republic and replaced the Queen with a president as the official head of state.

Much has changed in Mauritian politics, and yet much has remained the same. Since the Enlightenment, when the French colonized Ile de France, the island's political culture has thrived. The original European settlers educated their children, published newspapers and books, and supported the arts and sciences. Slavery and indentured labor, as well as racist attitudes toward Africans and Asians, ensured that until decolonization, people outside the European and assimilated Creole communities did not have full access to the public sphere. Some had access to community organizations, such as *baitkas* and cooperative societies, but as democracy took hold and literacy spread, non-Europeans came to support a plethora of schools, newspapers, publications, and political parties themselves. Mauritius became a healthy democracy, with more than 90 percent of registered voters casting ballots in regular elections. Many citizens are still skeptical about politics, particularly because of recent corruption scandals, but with the exception of the emergency of 1972–1976, there has not been any official repression of debate. This does not mean that there are no restrictions on the public sphere. On the contrary, the nature of local capitalism has limited the free flow of ideas, just as much as colonial politics did in the past. In fact, the two are deeply intertwined, as the sugar industry's involvement in politics demonstrates.

Between the 1810s and the 1970s, sugar production was the best response to the geographical and political constraints on the Mauritian economy. After independence in 1968, innovative economic planning, along with modern communications and finance, have enabled major economic changes to occur. Unlike other former sugar colonies, Mauritius did not become dependent upon European banks for credit. By the 1950s, the reverse was true. Bolstered by the marketing agreements, the Mauritian sugar estates became net investors in Britain and other European countries. The flight of capital from Mauritius only stopped when the sugar estates began to invest in local tourist hotels and textile manufacturing.[44]

In 1971, the government created an Export Processing Zone (EPZ), based on the Taiwanese model, which gave strong tax and other incentives to build factories in Mauritius. After 1979, when Mauritius implemented the IMF's structural adjustment program, the EPZ became very attractive for investors. Local commercial firms and sugar producers invested heavily, along with capitalists from Hong Kong and other countries. For the first time in 1985, Mauritius earned more money from exporting manufactured goods from the EPZ than it earned from exporting sugar. The EPZ produces mainly textiles from imported materials, and therefore still adds less value to exported goods than the indigenous sugar industry. The sugar marketing quotas also provide an important underpinning of finance capital for the EPZ.[45] But after independence, the government placed numerous restrictions on the sugar industry, including the mandatory year-round employment of workers and limits on factory closings. The industry would probably benefit from the same policies of liberalization that made the EPZ a success, but the historical legacy of *létam margoz* makes the sugar estates easy targets for politicians courting votes at election time.[46]

Nevertheless, the sugar industry continues to play a leading role in Mauritius. The non-sugar sectors of the economy have grown steadily, leading many to believe that manufacturing, tourism, and financial services are the waves of the future. And yet, any casual visitor can observe the continued importance of sugar cane on the landscape; people still cultivate it on almost every piece of arable land. From the air, the island appears flooded in a sea of green canes that recedes only at the edges of towns and mountains. This scene would not have been possible without intensive efforts on the part of scientists to improve the sugar cane plant itself.

The MSIRI during Decolonization and Independence

Between 1953 and 1992, debates about the MSIRI continued to focus on whether private industry or the government could manage a research institution most efficiently in the public interest. The continued importance of the sugar industry ensured that the MSIRI was a private institution with a major public role. But the colonial state and the Chamber of Agriculture created the MSIRI as a way of furthering the old style of corporatist government in the face of democratization and decolonization. While the undemocratic colonial regime was collapsing, this new private institution was intended to help preserve the undemocratic influence of technical experts while supporting the scientific needs of the old industrial elite. Not surprisingly, the MSIRI's close ties to the sugar industry's elite made it suspect in the eyes of many Mauritian farmers, who demanded government and small planter representation on the the institute's board. Even after the 1970s, when the government increased its power over the institute, the MSIRI still came under fire from some small planters and critics of the sugar industry.

During the 1950s and 1960s, the Chamber exercised a preponderant influence over the MSIRI. For the first few years, the institute's director, Wiehé, consulted with the Chamber on all the minute details of finance and administration. But by the 1960s, the institute staff was confident enough to handle the minutiae itself.[47] The Chamber still participated in the institute's larger decisions in two ways. On an informal level, the estate proprietors and staff members had a close working relationship with the scientists at the MSIRI and knew many of them through family ties and personal friendships. On a formal level, the Chamber had strong representation on the governing council of the MSIRI. The Chamber's delegate usually served as chairman, while the Chamber could rely upon the three representatives of the estates with factories and the one representative of the large planters without factories to share its views on most issues. The government supplied only one representative and the small planters two.

Beginning in the late 1960s, the MSIRI suffered from a shortage of funds, jeopardizing sugar cane research once again. The annual expenditures of the institute were fixed to enable a certain amount of long-term planning. But the annual income that derived from the levy on sugar exports varied according to sugar production. In bad years, perhaps with a cyclone or a drought, the MSIRI operated at a deficit. This hindered planning and reduced staff morale so much that in 1966, the institute asked the government to raise the duty on sugar.[48] The government did nothing, and in 1968, the institute reported that its financial situation had reached a "critical stage" and a "breaking point." The MSIRI argued that only seven-tenths of one percent of sugar export revenues supported its research, when comparable institutions in other countries received between 1.2 percent and 1.5 percent.[49] The government was unsympathetic and did little to help, except to change the collection of revenues from the Customs Department to the Sugar Syndicate, so that in addition to sugar exports, the small amount of sugar consumed in Mauritius was also subject to the tax.[50] In 1972, the government agreed to raise taxes on sugar to support the MSIRI, but only on the sugar that the estates with factories and large planters produced. In fact, the government took the opportunity to reduce the levy on small planters' sugar. The next year, rates for estates and large planters continued to rise, while small planters' taxes remained the same.

Reliance upon the government for additional finances forced the Chamber to relinquish some of its control over the MSIRI. The government raised its representation on the MSIRI council from one to three. Representatives of the ministries of agriculture, finance, and economic planning and development now joined the council, and their three votes, plus the two votes of the small planters, could balance the estates' five votes.[51] The Chamber's representative still retained the chairmanship, but after 1987, the government and small planters' representatives began to share that position as well. Ironically, just as the Chamber was losing its influence at the MSIRI, it was becoming more representative of the sugar industry as a whole. The estates still predominated in the organization, but small planters were starting to play a role, too.

During the 1980s, changing financial arrangements reduced the influence of the estates in the MSIRI even further. In 1984, the government established a parastatal organization called the Mauritius Sugar Authority, upon the recommendations of an economic commission. The Sugar Authority was supposed to monitor all aspects of sugar production and distribution, including research at the MSIRI, and to advise the Minister of Agriculture on how government policy should respond to the expressed desires of the industry's different sectors. The government assigned it the task of approving the MSIRI's expenditures, among many other duties. In reality, the Sugar Authority became an extension of the Ministry of Agriculture, adding another layer of government bureaucracy to the negotiations involved in setting the MSIRI budget and research agenda. During the late 1980s and early 1990s, the World Bank also played an important role at the MSIRI, which would not have been possible without government involvement. In 1986, the bank provided loans to improve the sugar industry's productivity, and between 1987 and 1990 it financed about one-half of the MSIRI's budget, providing an average of Rs.4.9 million annually, about $325,000.[52]

The contest between private interests and the state did not visibly hinder the MSIRI's research, because there was broad recognition that the government and the sugar industry had to cooperate. There was no question of returning to purely state-sponsored research, because the reasons for the transfer of the SRS to the MSIRI in 1953 still obtained. MSIRI salaries were significantly higher than those prevailing in the civil service, giving the best scientists little incentive to work for the government. On the other hand, if the sugar industry had tried to privatize research completely, it would have encountered difficulties in financing an institute without government support. The balance of power on the institute's board shifted somewhat from private industry to public officials, but industry-state collaboration proved fruitful, especially considering the alternatives.[53]

The MSIRI produced remarkable research results, and became one of the world's leading sugar cane research centers. In the early years, scientists there focused on the same problems that had faced the SRS, namely fertilizers and cultivation practices, pests and diseases of sugar cane, and factory technology. As time passed and new chemical and mechanical technologies became available, MSIRI agronomists also turned their attention to weed control, irrigation, and mechanization. Beginning in the late 1960s, the institute also began experiments with food crops that could be planted between rows of young canes and harvested before the canes grew too high and blocked the sun. The MSIRI was particularly successful with potatoes, maize, and tomatoes, helping to reduce the island's reliance upon outside sources for food. Beginning in the 1980s, the institute also embarked upon a program of economic and sociologic research, after criticism from the World Bank and others that scientists paid too little attention to the ways in which new agricultural technologies changed farmers' lives. Since 1961, outside analysts have consistently validated the MSIRI's success.[54]

The production of new sugar cane varieties remained at the forefront of the MSIRI's research program. Breeders at the institute provided the sugar industry with numerous new hybrid canes adapted to the island's various microclimates and soils. The MSIRI also imported several useful new varieties. (See Table 5.) Breeders produced so many new varieties that they essentially made the question of varietal decline a moot point. In the early 1990s, most estates were harvesting canes up to a sixth or seventh ratoon, and if the MSIRI could not provide that estate with a better variety after six or seven years, then its scientists believed that the breeding program was not living up to its standards. The program itself remained fundamentally similar to the SRS: each year breeders created tens of thousands of seedlings, of which one or two varieties were selected after a dozen years of trials. The primary criteria for selection have also remained the same: canes must grow vigorously, have a high sucrose content, and resist the island's principal diseases.

Beginning in the 1980s, MSIRI cane breeders also applied some new criteria to cane selection trials. For example, upright growth can facilitate mechanical harvesting, while canes with too many leaves add to an increasingly costly labor bill. The MSIRI started to provide varieties tailored to the island's principal microclimates, so that selection trials could also take into account tolerance to drought and super-humid conditions. In the future, other sugar cane traits may also become important. During the oil crisis of the 1970s, many sugar estates began to generate electricity from bagasse. This meant that canes with a higher fiber content became more economical to process than they were in the 1930s, when the factories rejected canes with high fiber content.[55] Since the 1930s, cyclone resistance has also been an important criterion for selection, and the canes with higher fiber content tend to resist cyclones the best.[56] Therefore, it is not necessarily such a bad thing that the average annual sucrose content of Mauritian sugar canes has not increased very much since the early 1960s. (See Table 6.) Environmentalism may also change cane breeding. According to the general secretary of the Chamber of Agriculture, markets may be developing for "organic" sugar cane grown without herbicides, fungicides, or artificial fertilizers and processed without chemical additives. Creating

Table 5

Principal Cane Varieties Cultivated in Mauritius, 1955-1990, Expressed as a Percentage of the Total Area under Cane on Sugar Estates

Cane	1955	1960	1965	1970	1975	1980	1985	1990
M134/32	74	25	5	1				
Ebène 1/37	15	26	11	1				
Barbados varieties	6	19	17	3				
M147/44		19	29	12				
M31/45		1	5	6	3	3	3	1
M202/46			11	12	3			
M93/48		1	12	21	12	9	7	1
M442/51			1	10	3			
M13/56				8	22	28	23	7
M377/56				8	13	18	4	
S17				4	31	23	9	5
M555/60						2	8	10
M574/62						2	8	1
M2173/63						1	7	3
M3035/66							6	23
R570							4	25

Note: "M" is the prefix used before the numbers of canes bred by the the Sugarcane Research Station, 1930–1953, and the Mauritius Sugar Industry Research Institute, 1954–Present. The number after the slash signifies the year in which the cane was bred.

Sources: Guy Rouillard, *Historique de la canne à sucre à l'île Maurice*. (Port Louis: MSM, 1990), 32. Mauritius Sugar Industry Research Institute, *Annual Report of 1990*, 79.

economical canes that can thrive without chemicals may indeed prove challenging.[57]

The MSIRI has had numerous successes, but to a certain extent it has failed the small planters. The institute worked closely with many estates in varietal and agronomical trials, but relied on the estates' agronomists and "small planter advisors" as well as on the Ministry of Agriculture's extension agents to convey information and materials to small planters. Most scientists at the MSIRI believed that any cane or technique designed for the estates would work in the fields of small planters, but small planter cultural operations can be quite different from the estates. Small planters generally practice longer ratooning, and the fragmentation of plots and smaller economies of scale make cultural practices relatively more expensive. As the island economy continues to diversify, increases in the cost of

Table 6

Sucrose Content of Mauritian Sugar Cane, 1954-1990

Year	Average Percentage of Sucrose in Cane (Factory Extraction	Average Percentage of Sucrose in Cane (Laboratory Extraction)	Year	Average Percentage of Sucrose in Cane (Factory Extraction)	Average Percentage of Sucrose in Cane (Laboratory Extraction)
1954	11.65	13.44	1973	11.51	13.05
1955	12.61	14.24	1974	11.68	13.26
1956	12.95	14.62	1975	10.85	12.56
1957	12.94	14.59	1976	10.78	12.36
1958	12.15	13.77	1977	11.05	12.74
1959	12.24	13.76	1978	10.68	12.28
1960	9.85	11.83	1979	10.90	12.63
1961	11.19	12.81	1980	10.42	12.10
1962	11.52	13.19	1981	10.84	12.48
1963	11.93	13.47	1982	10.45	11.95
1964	11.85	13.45	1983	11.51	13.16
1965	11.10	12.50	1984	11.49	13.97
1966	11.60	13.20	1985	11.57	13.03
1967	10.98	12.46	1986	11.73	13.21
1968	11.58	13.10	1987	11.09	12.65
1969	11.48	13.01	1988	11.50	13.18
1970	11.25	12.86	1989	10.45	12.06
1971	11.82	13.41	1990	11.25	12.85
1972	10.87	12.33			

Source: *President's Reports of the Mauritius Chamber of Agriculture,* 1954-1990.

labor exceed rises in the the price of sugar, and many small planters would be even more willing than the estates to sacrifice some sucrose content for canes that are more easily cultivated. In any case, small planter yields continued to be considerably lower than those of the estates.[58] (See Table 7.) To a certain extent, this disparity resulted from economies of scale. However, small planters never had the same access to science and technology as the estates had, because until the 1970s they lacked clout in planning the agendas of research institutions.

After 1937, the small planters' struggle against colonialism was deeply intertwined with their efforts to procure new canes and other forms of agricultural assistance. However, during the twenty years after 1937 the Department of Agriculture's extension policies changed on the surface while the services remained fundamentally similar. At first, the department rushed new cane varieties to the small planters, but the Second World War interrupted all of the department's activities, and extension officers found themselves called away to perform a wide variety of tasks related to the war effort. In 1947, a new Department of Cooperation took responsibility for the cooperative movement, removing a major distraction from the work of extension officers. In addition, the government directed the officers of the SRS to involve themselves in extension.[59] However, low salaries and staff shortages limited the department's ability to take full advantage of the situation. The success of M134/32 among the small planters filled the technological breach and helped to quell agrarian unrest, with the notable exception of the riots on some northern estates in 1943. However, without an expansion of extension and research services for small planters, distributing M134/32 could only be a temporary measure.

In 1957, the Department of Agriculture became the responsibility of a cabinet minister, and in the years that followed the number of extension agents increased dramatically as the government sought to reach small planters. Each of the island's seven rural districts received a senior extension officer and several junior officers, depending on the size of the district and the number of sugar factories. The more urban districts of Port Louis and Plaines Wilhems shared access to the extension offices of the Pamplemousses and Black River districts, respectively. Even with this expansion of extension services, and the enthusiasm of many of the officers, it remained difficult to reach many small planters. The ratio of officers to farmers has always been low, and many small planters worked other full-time jobs, making them difficult to locate. In the late 1980s, 20 percent of all small planters were women, but the Ministry of Agriculture was not hiring female extension officers and had no special programs for women.[60] The Ministry also was not providing extension officers with adequate funds to purchase gasoline, limiting their ability to visit farmers in the fields. Extension officers have always devoted a great deal of their time to instructing farmers in food crop cultivation because of the government's effort to make the island more self-sufficient. More fruits and vegetables became available after independence, but small-scale sugar cane planters needed attention too. In 1971, the Chamber of Agriculture's survey of small planters revealed that many of them had never even heard of the Ministry of Agriculture's extension services.[61]

During the 1970s, a number of Mauritians expressed dissatisfaction with the state of small planter extension services, initiating a complete

Table 7

Metric Tons of Cane Harvested per Hectare
Estates with Factories, Owner-Planters and Tenant-Planters, 1953–1990

Year	Estates with Factories	Tenant- Planters	Owner- Planters*	Island Average
1953	77.6	47.1	57.1	66.2
1954	73.8	38.1	50.0	60.5
1955	74.0	33.8	48.3	59.7
1956	76.2	38.8	51.2	62.6
1957	76.6	34.3	46.6	60.9
1958	73.1	36.2	45.0	58.3
1959	77.4	37.6	47.8	61.6
1960	36.4	21.2	24.5	30.2
1961	76.6	32.8	50.0	62.8
1962	66.6	40.0	46.9	56.9
1963	83.5	42.6	57.6	70.4
1964	62.4	33.8	44.7	53.3
1965	85.0	46.4	61.2	61.2
1966	70.2	37.6	47.1	58.8
1967	84.0	50.2	59.5	72.1
1968	74.3	42.1	55.2	64.7
1969	85.9	45.5	61.4	73.5
1970	75.0	41.2	52.6	64.0
1971	80.2	44.7	50.0	66.2
1972	90.4	55.5	66.6	79.0
1973	88.8	53.6	65.2	77.4
1974	88.3	59.3	48.8	75.0
1975	63.5	34.3	43.3	54.3
1976	89.5	56.0	67.8	79.5
1977	85.0	53.3	63.5	75.0
1978	87.8	53.6	67.4	78.3
1979	88.8	53.6	68.5	79.5
1980	64.3	38.6	50.2	57.8
1981	78.1	49.0	55.2	68.1
1982	91.4	57.6	73.1	82.8
1983	76.6	48.1	56.9	67.4
1984	75.4	43.8	52.6	64.5
1985	80.4	50.7	63.1	71.9
1986	86.4	53.6	70.0	77.8
1987	89.3	56.2	72.8	80.7
1988	85.2	48.6	59.3	72.1
1989	81.2	43.8	61.4	71.2
1990	85.9	42.8	59.6	72.7

* Includes small planters as well as large planters without factories.

Source: *President's Reports of the Mauritius Chamber of Agriculture*, 1953–1991.

reconsideration of extension methods. In 1973, a government commission of enquiry blamed inadequate extension services in part for the continuing yield gap.[62] The Chamber proposed that the MSIRI develop its own extension service for small planters, rather than rely on passing along information indirectly through estate agronomists and government extension officers. In the late 1940s, the SRS had already tried this direct approach, but despite some successes the project became bogged down in the SRS's institutional problems. The Chamber also encouraged the estates to make stronger efforts to advise small planters, but most estates responded half-heartedly.[63] In 1980, the Chamber changed its tack, recognizing that the fragmentation of small planter landholdings placed serious limitations upon traditional extension efforts. It suggested that unifying neighboring small planters into larger "management units" would increase the likelihood of small planters adopting new methods.[64]

In 1985, the government and the sugar industry cooperated in creating an experimental extension organization for small cane planters called the Farmers' Service Centres (FSCs). These have changed extension services considerably in four sugar estate factory areas where the experimental "pilot project" FSCs were located when I visited them in 1992. The FSCs are run as a private corporation under the supervision of the Sugar Authority and they receive financing from the World Bank. One extension officer seconded from the Ministry of Agriculture manages each FSC, together with about six assistants. The FSCs act as a kind of technology supermarket for small cane planters. Representatives of merchant firms sell fertilizers, herbicides, and other products at the FSCs, and small planters can also arrange to hire tractors and transport there. The FSCs have encouraged some farmers on adjacent lands to join themselves into Land Area Management Units (LAMUs), facilitating mechanical operations. The MSIRI has a special liaison officer with the FSCs, who assists in disseminating the institute's research findings to the small planters and who has arranged for them to visit the MSIRI and offer their thoughts to the institute's cane breeders.

Given the history of witnessing as a technique for establishing the credibility of new canes, it is highly significant that each FSC also has encouraged some of its most reliable farmers to cultivate cane nurseries. In fact, the FSCs have identified a number of the best nearby farmers and enlisted them in the extension effort as "contact farmers" charged with informing their neighbors about the latest techniques. The FSCs also identify the 200 lowest-yielding farmers in each area and provide them with intensive assistance. The FSC extension officers supervise all aspects of this holistic approach to ascertain that farmers make the best possible decisions. The FSCs have generally been successful, despite some minor problems, and their extension officers have reported that some small planters have achieved higher cane yields than nearby estates. In 1993, four more factory areas acquired them.[65] Nevertheless, the FSCs still face an uphill battle in convincing small planters with disparate interests to cooperate in managing their lands.

In 1937, the colonial government realized that it could use new cane varieties and extension services as part of its strategy to quell anti-colonial agitation among the small planters. The advent of parliamentary democracy changed the nature of the state sufficiently so that beginning in 1957 the increasingly independent government saw the provision of technology and services to the small planters as a way of maintaining good relations with constituents. The FSCs are potentially an improvement over traditional extension methods, but the Sugar Authority that supervises them is not as independent of the government as it was supposed to be. The FSCs' connections with external aid organizations such as the World Bank may grant them some leverage in their dealings with the Ministry of Agriculture, but the World Bank's demands may not always be consistently in the interest of all small planter constituencies.

State pressure and electoral politics may make it difficult to institute research and extension services for small planters. The owners of the large estates complain that the transfer of political power to the small planters has given the least efficient sector of the sugar industry the most subsidization. The Ministry of Agriculture has also imposed export duties and labor laws on the large estates that give small planters special advantages. The government has also delayed or prevented the closure of several inefficient factories. Democratic politicians exacting revenge on the factory-estates for the sins of the past may not necessarily bring the most economic benefits to the island. Each side of the debates about sugar cane science and technology continues to define the public interest in order to enhance its own authority.

During the late 1980s, the MSIRI began to dispense with the mediation of the state in establishing credibility with small planters. The MSIRI invited groups of small planters to tour its facilities and to meet with cane breeders. This approach to constituents harkened back to the days when the Chamber's delegates inspected the government's cane collection at the Pamplemousses Gardens. A scientific institution's security is deeply intertwined with the way in which it persuades its patrons that it is effective. Devising new ways of witnessing experiments will not only help to make research institutions more democratic. It will also give the witnesses and the people they represent a stake in the scientific institution.

The MSIRI has earned a strong international reputation, based on its numerous scientific accomplishments, but some small planters accuse it of catering almost entirely to the needs of the estates. This is a matter of perception, closely related to the way that sugar research established its credibility. The biggest sugar estates argue that it is impossible for research institutions to produce a cane that is good for the estates but bad for the small planters. But in 1953 the colonial government did collaborate with representatives of the estate owners to found the institute. In fact, the MSIRI is the embodiment of a century of cooperation between the government and the estates' main representative organization, the Chamber of Agriculture. The MSIRI pledges to serve all sectors of the Mauritian sugar economy, but it has failed the small planters to a certain extent. In recent years, the institute has attempted to correct these problems, but it is difficult to disentangle culture and politics from the production of better canes.

The current institutional arrangements for the production and distribution of new plants may allow Mauritians to adjust well to future changes in sugar cane science. Competition between the estate sector and the small planter sector will increase, particularly if the cost of labor continues to rise. The sugar industry also continues to depend on the good graces of Europe for a market, which is a tricky proposition at a time when there is stiff competition from other sweeteners such as high-fructose corn syrup and aspartame. New approaches to extension, such as the Farmers' Service Centers, have the potential to broaden access to new technology and information so long as the MSIRI's research program remains healthy. By 2010, it is likely that a map of the sugar cane genome will exist, enabling scientists to improve the plant through genetic engineering. But cane breeders will not have a monopoly on biotechnological methods: these might help producers of rival sweeteners, too. At the very least, the Mauritian institutions that produce and distribute plants are structured in such a way that they are unlikely to allow one sector of the sugar economy to monopolize new knowledge and use it to squeeze out other producers. The future holds numerous challenges, but Mauritius is in a good position to meet them. Over the course of the past 150 years, there has been a progressive incorporation of planters' knowledge within institutional research programs. Initially, small planters were excluded from this process, but gradually they are being incorporated into the research agenda of the MSIRI.

CONCLUSION

CREOLES AND HYBRIDS

Mauritius is filled with hybrids, as anyone can see. Hybrid canes cover the landscape, demonstrating the endurance of the sugar industry and the success of the cane breeders. But hybrid sugar canes are more than just productive plants. They embody the politics of Mauritian culture.

Mauritians themselves are hybrids. There are no native Mauritians, but instead the islanders come from many different backgrounds. This is not to say that Mauritians do not associate themselves with ethnic identities. They do, like people in the rest of the modern world. Mauritians identify themselves as Franco-Mauritians, Indo-Mauritians, and Sino-Mauritians, and also as Creoles and Muslims, and they even belong to various subsets of all these groups. National identity is also important in Mauritius. Some islanders promoted ethnic and national identities as part of the struggle to overcome colonialism. Later, some islanders promoted national identity and class identity as a way to transcend these same ethnic identities.

But in Mauritius, ethnic identity, national identity, and class identity have never coalesced fully. Instead, they are subject to continuous negotiation. Thomas Eriksen points out that even the Kreol word for nation (*nasyon*) has many different meanings: it can indicate an identification with the nation-state, but it can also mean caste, ethnic group, race, or linguistic community. Such identities are not necessarily exclusive, but depend on the context in which they operate.[1]

Mauritian life is an object lesson in what W.E.B. DuBois called "double consciousness." In a book on African-American intellectuals, Paul Gilroy argues that W.E.B. Du Bois, Richard Wright, and other writers resolved the inherent contradiction between ethnic and national allegiance by accepting that they inherited a culture that was a "creolized" and "hybridized" mixture of transnational and multicultural influences.[2] The same broad pattern is true in Mauritius: the continuous creolization of Mauritian culture renders the boundaries around these identities fundamentally ambiguous. As Eriksen argues, one can only understand Mauritian ethnicity when one understands its fluidity: that it is invoked in particular, varying circumstances, and that it has constantly shifted over time.

The Kreol language itself illustrates this point very well. It is the language that the African slaves made while working on the French plantations, and it blends together elements of African languages with French. But today Kreol is not just the language of the slaves' descendants. Over the course of the nineteenth and twentieth centuries, Hindu, Muslim, and Chinese immigrants all came to speak it as either their first or second language. Vinesh Hookoomsing has shown that even as Indo-Mauritians preserved ties with India, every census indicates that fewer and fewer speak Indian languages. Speaking Kreol is the mark of acculturation in Mauritius; it is not to be confused with efforts on the part of some Creoles to articulate a separate identity, but it is to be associated with the making of an island culture derived from many origins.[3] The orthographic distinction between Kreol and Creole itself represents consciousness of multiple identities, but the word "creole" also implies change. The Oxford English Dictionary says that the first people to be called Creoles were Africans and Europeans born, raised, or naturalized in the Caribbean. To be creole is not just to have a creole identity, but to participate in a process of creolization.

Mauritian identity is filled with ambiguities and surprises, and the processes of cultural creolization, hybridization, and acculturation continue. For example, even though snobs reject Kreol as a low-class language, some elite Franco-Mauritians speak Kreol with each other, especially when they are feeling loose and telling jokes. And yet, as Larry Bowman has shown, when the radical MMM government of 1982–83 tried to make Kreol the national language, the party met with resistance from most Mauritians. Some consider Kreol to be an inferior language, even though they speak it in their homes. Others fear that Kreol is not a recognized international language, and that if Mauritius has to depend on the outside world for its economic and political security, it is better to know English and French.[4]

The Kreol language shows the complexities of cultural identity in

Mauritius, but there is no better symbol for the creolization of culture than the sugar cane plant itself. Farmers and scientists imported it from overseas, then they adapted it to local circumstances by crossing it with other imports. But this process of adaptation was not only about finding which new canes yielded the most sugar in the peculiar soil and climate of Mauritius. Sugar canes were associated with significant cultural, economic, and political practices, and scientists had to make the new canes credible to the different cultures and classes that are present on the island. Decisions about canes reflected and influenced the construction of cultures and states in Mauritius.

In Mauritius, as in other states, notions of political sovereignty bore a close relationship to the constitution of scientific credibility. Modern states founded their sovereignty upon the public witnessing of political events. In autocracies as far removed as the France of Louis XIV and the China of Mao Zedong, states incorporated citizens into the rituals of statecraft. In democracies, people could participate in government by voting and could experience politics indirectly through the media. But there are always limits to public participation. Governments consider it impractical for every member of the population to participate directly in governance. Therefore, they design methods of virtual participation. For example, in democracies people delegate authority to legislators and the news media.

Likewise, modern scientists founded their credibility upon the virtual witnessing of experimentation. In theory, laboratories are open to public scrutiny. In practice, they are closed to the public and open only to professional scientists. Then how do scientists establish their credibility with people outside the laboratory? Scientific witnesses report to the public on experimental successes and failures. Their reports appear in refereed professional journals as well as in the news media. Like modern states, modern scientists rely on witnesses to represent their activities.[5]

But there is more than just a parallel between the use of witnessing in modern states and sciences. The British colonial government of Mauritius derived most of its support from a tax on sugar exports. The production of sugar supported the colonial state, and it also supported the elite owners of sugar estates and factories. Sugar production also gave structure to the lives of lower-class Mauritians, whether they were field workers, domestic workers, or dock workers. Sugar was sovereign in Mauritius. The sugar cane plant did not determine Mauritian history; it covered the landscape because Mauritians and outsiders consented to support it. Among the team of planters, officials, and others who manufactured consent about sugar, one could also find the scientists who ensured that sugar canes would survive.

When the sugar economy declined during late nineteenth century, planters recognized how important it was to obtain new, higher-yielding, disease-resistant cane varieties. They turned to cane collectors and government botanists to provide better varieties. Beginning in the late 1860s, the Chamber of Agriculture organized expeditions to search the South Pacific for previously unknown varieties. They gave the canes to the state botanists at the Royal Botanic Gardens in Pamplemousses, expecting them to care for them and to select the best ones for distribution. But the Franco-Mauritian planters did not completely trust the British gardeners. Meller, Horne, and their associates were not only from a different culture, but they were also more interested in botany than in agriculture. The Franco-Mauritian planters had a rich tradition of activism and research in the agricultural sciences, and they found a way to supervise the work of the gardeners: they arranged for virtual witnesses to visit the gardens and scrutinize the work of the state gardeners.

This kind of planter influence over a state institution was not at all unusual. The sugar barons were creolizing the gardens, just as they were creolizing the colonial state. This was not at all unusual for a British Crown Colony, where relatively weak states often shared power with local elites.[6] Such corporatist colonial governments could enroll local scientists into state objectives, while at the same time scientists could enroll the state to attain their own objectives. The Mauritian elite had a long local scientific tradition, which began to influence the production of knowledge in the colonial department of agriculture. At the same time, it never occurred to the Chamber that the burgeoning classes of Indian small planters might desire a say in the debates about sugar cane research institutions. Indian immigrants were well on their way to purchasing one-third of the lands under cane, but they were not yet fully acculturated to Mauritius, and they did not yet have the power to supervise botanical research.

The Chamber took a proprietary attitude toward the state gardens at Pamplemousses, but in the end it was not satisfied with their performance. The obligations of the state gardeners to perform other duties mandated by Kew Gardens and the local government, conflict between the gardeners and the Chamber, and the shift in European agricultural sciences toward research laboratories, all persuaded the Chamber to create a private experiment station of its own. In 1892, the Chamber played an instrumental role in founding the Station Agronomique, where Philippe Bonâme studied plant breeding, soil chemistry, factory chemistry, and plant pathology, among other things. Bonâme accomplished a great deal, but he failed to provide the sugar barons with productive seedling canes.

In 1913, the British colonial government and the Chamber of Agriculture agreed to incorporate the Station Agronomique into a new government Department of Agriculture on the model of the West Indies. The department was founded to provide a full range of scientific services, including plant breeding, entomology, and extension to all farmers. The Chamber supervised the work of the department, but the story of the department's staffing shows the process of how a government institution could be creolized. Many department officials were British expatriates, but many were also Mauritians. At first, the department hired autodidacts from the Franco-Mauritian scientific tradition, like Donald d'Emmerez de Charmoy and Charlie O'Connor. The department made it possible for Mauritians to receive recognized training in the agricultural sciences at Réduit, and soon it began to hire Mauritians who had trained there. These Mauritians worked side by side with their British counterparts, and some made significant contributions to the agricultural sciences: Charlie O'Connor and Alfred North-Coombes in agronomy; Donald d'Emmerez de Charmoy, André Moutia, and Jean Vinson in entomology; Gabriel Orian in pathology; and perhaps most importantly, Aimé de Sornay in plant breeding.

Even as these Mauritians received increasing amounts of recognized, European-style training in the agricultural sciences, they also creolized the colonial department of agriculture. It is a classic story of cross-enrollment: the British sought assistance from Mauritians and trained them according to metropolitan and imperial standards, but this helped Mauritians to exert greater influence upon colonial agricultural research. It is also significant that none of the people mentioned above was an Indo-Mauritian. Even as Indo-Mauritians and small planters were having more say in the governance of Mauritius, they were not yet enrolled in the project of agricultural research.

Indo-Mauritian small planters had been processing knowledge about sugar canes since the *grand morcellement* of the late nineteenth century. For decades they had lurked on the margins of colonial science. Then after 1937, the colonial state disseminated new sugar canes and agricultural extension services to small planters as part of an overall strategy of social control. In 1948, Ramgoolam and the Labour Party began to add the voices of non-elite Mauritians to decisions about the production and distribution of sugar cane science and technology. As decolonization gathered steam during the 1950s and 1960s, non-elite Mauritians had more of a say in their own government. Now that the small planters carried some political clout, they could participate in the politics of producing knowledge about the sugar cane.

The credibility of both the new Mauritian state and sugar cane science were closely related. During decolonization, the Mauritian state preserved many of its corporatist attributes while enrolling the vast majority of the population into participation through voting and through a highly politicized press. At the same time, Mauritian sugar cane research institutions preserved their research agendas while incorporating the input and witnessing of the representatives of the small planters. On a practical level, sugar cane research remained fundamental to the island's economic and social order. Even the textile and tourist industries that dominated the economy by the 1980s relied heavily upon the sugar industry as a stable source of capital. On a more theoretical level, the small planters and the wider population adopted new notions that they had a right to supervise the working of institutions that affected their lives in important ways. The response of sugar cane research institutions to democratization mirrored the response of the state as a whole. Representatives of the small planters would be incorporated to witness sugar cane research and report to their consituencies, while the basic outline of research and extension services did not change very much at all.

In a sense, this history confirms Robert Chambers's position that the best agricultural science incorporates the knowledge of the farmers themselves.[7] But getting farmers to participate in agricultural science is not just a question of granting them an opportunity to influence research agendas and to provide tidbits of productive knowledge, as crucial as this influence and information may be. It is also a question of culture and credibility. In colonial Mauritius, metropolitan and local scientists became embroiled in cultural, economic, and political disputes. Scientists made their work believable, not only because their canes yielded more sucrose, but also because the right people witnessed the canes and granted their assent. The experimental landscape shifted repeatedly, from the Pamplemousses Gardens to the Station Agronomique to the Department of Agriculture to the SRS and finally to the MSIRI, but in each case scientists depended on the state and the producers and granted them the authority to supervise their work. At the same time, some Franco-Mauritians and Creoles preserved their tradition of amateur cane breeding, while others joined the SRS and the MSIRI. Indo-Mauritians made a contribution, too, by selecting and distributing the Uba canes.

Such arguments about the relations between culture, politics, and science are bound to be controversial, and it is important to be aware of some counter-arguments. The role of science in agriculture is most commonly understood from the perspective of supposedly rational choices. For example, it might be argued that farmers adopt new sugar canes simply because they give higher yields. Recently, proponents of such "realistic," "common-sense" interpretations of science have attacked advocates of a cultural approach to the study of science. But if the story of the sugar cane is a story of the unfolding of obvious scientific advances, then why did the production and distribution of better plants involve so much cultural, economic, political, and social maneuvering?

In Mauritius, colonial agricultural scientists established their credibility first and foremost among the wealthiest plantation owners. These elite farmers gained the power to influence the research agendas of imperial scientists "on the spot." Sometimes small planters selected and distributed their own plants, depending on their relations with the elite. At first, colonial scientists largely ignored the needs of small planters. They paid more attention to small planters after the 1937 riots, when they promoted their own knowledge to small planters mainly to help the colonial government control rural rebellion.

This story stands in contrast to some of the rational-choice accounts of plant-breeding in today's former colonies. In his book on modern varieties of food crops, Michael Lipton argues that by and large, modern plant-breeding has had a positive impact on the lives of poor farmers. He acknowledges that breeders and farmers do not necessarily share the same objectives: breeders tend to concentrate on improving yields while poor farmers also seek sustainability. Farmers may not get exactly what they want from breeders, but according to Lipton, "Common sense . . . counts. More food is likely to mean more saved lives . . ."[8]

Nevertheless, this book has shown that there is more to the story of canes in colonial Mauritius than rational-choice analysis can take into account. The new canes may have been demonstrably better, but as the phrase implies, they had to be demonstrated as such. The very debates about new varieties tell a story: that plant breeding was closely bound up with the power relations of a colonial society. Plant breeding in colonial Mauritius can only be understood in the context of colonial culture and politics.

The story of sugar cane is also not just a tale of economic or political groups trying to capture the production of new plants in order to serve their own interests. It is a story of how the sugar cane plant took its place in many parts of the colonial landscape, permanently altering ecological and social relationships. It is a story of how people produced knowledge about the plant in many different locations, some of them unexpected and for-gotten. It is also a story of how different producers of knowledge opened and closed their doors to public participation, persuading each other to adopt new practices.

The rise and fall of different groups of Mauritians to political and economic authority can be seen in the history of supervision and witness-ing at the sites for producing and distributing sugar canes. The history of Mauritius demonstrates that farmers from many cultures and classes have shaped the production and distribution of science and technology. Con-ventional histories of "technology-transfer" have ignored these farmers, choosing to focus instead on metropolitan states and scientific institutions. Mauritian history suggests that metropolitan European interests were not quite so powerful in setting the research agendas of colonial scientific insti-tutions. The case of Mauritius suggests that anyone hoping to remedy the state of the global environment or to create sustainable food production in former colonies must expect to encounter a complex cultural legacy. Even in the most remote places, even among the most downtrodden people, new kinds of science and technology are subject to conflict and negotiation.

NOTES

Notes to the Introduction

1. Shapin, *Scientific Revolution*, 9–10.
2. Fairhead and Leach, *Misreading the African Landscape*.
3. Schama, *Landscape and Memory*, 24.
4. Richards, *Indigenous Agricultural Revolution*.
5. Fitzgerald, *Business of Breeding*, 3.
6. Latour, *Science in Action* and *We Have Never Been Modern*.
7. Hughes, *Networks of Power*.
8. Law, "Technology and Heterogeneous Engineering," 111–14.
9. Barnes, *The Sugar Cane*, 91–102.
10. Brandes and Sartoris, "Sugarcane," 572–3.
11. Stevenson, *Genetics and Breeding*, 3–4.
12. Stevenson, *Genetics and Breeding*, 41.
13. Stevenson, *Genetics and Breeding*, 42.
14. Curtin, *Plantation Complex*.
15. Blackburn, *Sugar-Cane*, 74–5.
16. Blume, *Geography of Sugar Cane*.

Notes to Chapter One

1. Twain, *More Tramps Abroad*, 430.
2. Padya, *Weather and Climate*. Ramdin, *Mauritius*, 19–22.
3. Padya, *Weather and Climate*, 156–63.
4. A. North-Coombes, *Découverte*, 31, 73, 99, 110–31.
5. Chan Low, "T'Eylandt." A. North-Coombes, *Leguat*, 5.
6. Pitot, *T'Eylandt*, 71, 79, 81, 83–7, 232, 244–5.
7. De Nettancourt, "Peuplement," 220–24. Verin, *Maurice*, 93–4.
8. Toussaint, *History of Mauritius*, 21–3.
9. A. North-Coombes, *Leguat*, 8. Grove, *Green Imperialism*, 129–45.

10. Toussaint, *History of Mauritius*, 22–3.

11. Wanquet, "Café," 55–73.

12. Lagesse, *L'Ile de France*, 21–3.

13. Lagesse, *L'Ile de France*, 37.

14. Filliot, *La traite des esclaves*, 56.

15. Baker, *Kreol*, 1. Hookoomsing, "L'emploi," 76–88.

16. Filliot, *La traite des esclaves*, 56–7.

17. Toussaint, *History of Mauritius*, 30.

18. Filliot, *La traite des esclaves*, 58–9.

19. Ryckebusch, *La Bourdonnais*, 48–9. Grant, *Mauritius*.

20. A. North-Coombes, *History of Sugar Production*, 1–15.

21. Toussaint, *Histoire des îles Mascareignes*.

22. Grove, *Green Imperialism*, 190–91.

23. Grove, *Green Imperialism*, 199–254.

24. Alpers, "French Slave Trade." Filliot, *La traite des esclaves*, 51, 62, 69. Reddi, "Aspects of Slavery," 106. Gerbeau, "Slave Trade." Toussaint, *History of Mauritius*, 38–9, 52.

25. Allen, "Creoles," 35, 44.

26. Allen, "Creoles," 97–8.

27. Allen, "Creoles," 111.

28. Filliot, *La traite des esclaves*, 65–7.

29. Toussaint, *History of Mauritius*, 41–56.

30. Stevenson, *Genetics and Breeding*, 41.

31. Deerr, *Cane Sugar*, 47.

32. See Ly-Tio-Fane, *Mauritius and the Spice Trade, Notice historique*, and *The Triumph of Jean-Nicolas Céré*.

33. *Science* 156 (5 May 1967): 611–22.

34. Pyenson, *Civilizing Mission*, 3, 127.

35. For a more extensive critique of Pyenson, see Worboys and Palladino, "Science and Imperialism."

36. McClellan, *Colonialism and Science*, 289–90.

37. Grove, *Green Imperialism*.

38. A. North-Coombes, *History of Sugar Production*, 13–15.

39. Deerr, *Cane Sugar*, 47.

40. Deerr, *Cane Sugar*, 48–9.

41. Richards, *Indigenous Agricultural Revolution*.

42. Bartholomew, "Modern Science in Japan."

43. Toussaint, *History of Mauritius*, 53–4.

44. Graham, *Great Britain in the Indian Ocean.*

45. Graham, *Great Britain in the Indian Ocean.*

46. Burroughs, "Mauritius Rebellion," 244–5.

47. Bayly, *Imperial Meridian*, 193–216.

48. A. North-Coombes, *Evolution of Sugarcane Culture*, 165. Deerr, *History of Sugar*, 1:185; 2:531. MCOA, *Mauritius Chamber of Agriculture, 1853–1953.* For conversions to metric and discussion, see Storey, "Biotechnology and Power," 57, 60, 75.

49. *Mauritius Chamber of Agriculture, 1853–1953.* Lamusse, "Development in Field and Factory," 22–5. Galloway, *Sugar Cane Industry*, 134–41. A. North-Coombes, *Evolution of Sugarcane Culture*, 165.

50. Ly-Tio-Fane, *Notice historique*, 4–5, 8–10. Rouillard, *Historique de la canne à sucre*, 15, 38. Desjardins, *Telfair.*

51. Lamusse, "Sources of Capital," 355–6.

52. Baker and Hookoomsing, *Diksyoner kreol morisyen.*

53. Nwulia, *History of Slavery*, 42–6. Reddi, "Aspects of Slavery," 107–8. Carter and Gerbeau, "Covert Slaves," 194–208.

54. Nwulia, *History of Slavery*, 90–91.

55. Nwulia, *History of Slavery*, 126.

56. Mathur, "Partly-Elective Legislature," 74–5.

57. Burroughs, "Mauritius Rebellion."

58. M.D. North-Coombes, "Slavery to Indenture," 80.

59. Nwulia, *History of Slavery*, 144–9, 158–9, 173–8.

60. Allen, "Creoles," 159, 169–73, 180–87.

61. M.D. North-Coombes, "Slavery to Indenture," 84–5.

62. Reddi, "Establishment of Indian Indenture," 5–7.

63. Tinker, *A New System of Slavery*, 64–70.

64. Tinker, *New System of Slavery*, 72–5.

65. Tinker, *New System of Slavery*, 75–6.

66. Kuczynski, *Demographic Survey*, 2:793–8.

67. Northrup, *Indentured Labor*, 17–29, 59–70, 80–89.

68. Carter, *Voices from Indenture*, 62–92.

69. Tinker, *New System of Slavery*, 82–90.

70. Tinker, *New System of Slavery*, 106–7.

71. M.D. North-Coombes, "Slavery to Indenture," 98–100.

72. Tinker, *New System of Slavery*, ch.6.

73. M.D. North-Coombes, "Slavery to Indenture," 92.

74. Tinker, *New System of Slavery*, 39–60. Reddi, "Political Aware-ness," 1–3.

75. Reddi, "Labour Protest."

76. Lamusse, "Sources of Capital," 356–7.

77. Deerr, *History of Sugar*, v.2, ch.27.

78. Lamusse, "Sources of Capital," 359–60.

79. Stevenson, *Genetics and Breeding*, 42.

80. Ly-Tio-Fane, *Notice historique*, 8–10.

81. Bouton, *Rapport*, 12.

82. Rouillard, *Historique de la canne à sucre*, 15–16.

83. Bouton, *Rapport*.

84. Rouillard, *Historique de la canne à sucre*, 16.

85. Ly-Tio-Fane, *Notice historique*, 11.

86. Toussaint, *Chambre d'Agriculture*, 4–5.

87. Toussaint, *Chambre d'Agriculture*, 4–5.

Notes to Chapter Two

1. Toussaint, *History of Mauritius*, 71–4. *Mauritius Chamber of Agriculture, 1853–1953*. Deerr, *History of Sugar*, 2:531.

2. Paturau, *Histoire économique*, 160. Lamusse, "Mauritius Sugar in World Trade," 13, 24–5.

3. Toussaint, *History of Mauritius*, 75. Boyd, *Malariology*, 660, 714, 731, 803, 806.

4. Lamusse, "Development in Field and Factory," 26–7.

5. Galloway, *Sugar Cane Industry*, 134–41.

6. Lamusse, "Sources of Capital," 363–5.

7. Lamusse, "Sources of Capital," 363–5.

8. Bowman, *Mauritius*, 18.

9. Allen, "Slender, Sweet Thread," 182.

10. Allen, "Creoles," 211–24.

11. MSIRI, *Annual Report*, 1989, 16–19.

12. Allen, "Creoles," 231–9.

13. Carter, *Lakshmi's Legacy*.

14. Berry, *Fathers Work for Their Sons*.

15. Koenig, *Agricultural Census, 1930*.

16. Benedict, *Indians in a Plural Society*, 69–83.

17. Chamber of Agriculture (MCOA), President's Report, 1855.

18. MCOA, President's Report of 1856, 1.

19. MCOA, President's Reports of 1857, 5–6; 1858, 73–4, 90.

20. MCOA, President's Report of 1860, 13.

21. MCOA, President's Reports of 1861, 43; 1862, 8.

22. Bouton, *Rapport*.

23. Barnes, *The Sugar-Cane*, 34–7. Stevenson, *Genetics and Breeding*, 3.

24. MCOA, President's Report of 1862, 11–12.

25. Rouillard, *Le jardin des Pamplemousses*, 38.

26. Barkly to Colonial Office, ca.1864, 19–23, RBGK-3.

27. Rouillard, *Historique de la canne à sucre*, 15; *Le jardin des Pamplemousses*, 38.

28. MCOA, President's Report of 1862, 11–12, 104.

29. MCOA, President's Report of 1862, 105.

30. MCOA, President's Report of 1862, 106.

31. Duncan to Colonial Secy., 22 Sept. 1860; C.O. to Hooker, 4 Jan. 1861; C.O. to Hooker, 27 April 1861; Stephenson to Newcastle, 4 March 1861; C.O. to Hooker, 24 June 1861, 17, RBGK-3.

32. Grove, *Green Imperialism*. Drayton, "Imperial Science."

33. Curtin, *Death by Migration*.

34. Brockway, *Science and Colonial Expansion*, 113–21.

35. Drayton, "Imperial Science," 344.

36. Toussaint, *Dictionary of Mauritian Biography*, 366.

37. Barkly to Cardwell, 5 May 1865, 33–4; draft of letter of Hooker to the Colonial Office, 10 July 1865, 35; Royal Botanic Gardens, Pamplemousses (RBGP), *Annual Report*, 1865; all in RBGK-3.

38. RBGP, *Annual Report*, 1865; Horne to Col. Secy., 23 Feb. 1865, 199–205, RBGK-3. See also Duncan, *Catalogue of Plants*.

39. RBGP, *Annual Report*, 1866.

40. MCOA, President's Report of 1862, 125–6.

41. MCOA, President's Report of 1862, 124.

42. MCOA, President's Report of 1866, 4.

43. Rouillard, *Historique de la canne à sucre*, 18, 38.

44. MCOA, President's Report of 1866, 4.

45. MCOA, President's Report of 1867, 1.

46. RBGP, *Annual Report*, 1867.

47. Horne to Hooker, 5 Sept. 1867, Corr. 188.293, RBGK.

48. RBGP, *Annual Report*, 1867.

49. RBGP, *Annual Report*, 1868.

50. RBGP, *Annual Report*, 1868. Horne to Hooker, 17 July 1868, Corr. 188, RBGK.

51. MCOA, Minutes of 16 June 1868, 3, 22–3.

52. MCOA, Minutes of 1 Dec. 1868, 58.

53. Toussaint, *Dictionary of Mauritian Biography*, 1042–3.

54. MCOA, Minutes of 22 Sept. 1868, 38–9.

55. MCOA, Minutes of 22 Sept. 1868, 39–41.

56. MCOA, Minutes of 29 Oct. 1868, 49.

57. Brockway, *Science and Colonial Expansion*, 7.

58. Worboys and Palladino, "Science and Imperialism," 91–2.

59. Kloppenburg, *First the Seed*.

60. Berman, *Control and Crisis in Colonial Kenya*, 1–9.

61. Headrick, *Tentacles of Progress*, 107.

62. MacLeod, "Passages in Imperial Science," 123–7.

63. MCOA, Minutes of 29 Oct. 1868, 50–51.

64. MCOA, Minutes of 29 Oct. 1868, 50–51.

65. MCOA, Minutes of 1 Dec. 1868, 58.

66. Biagioli, *Galileo, Courtier*, 17.

67. Dear, "Totius in Verba," 155–7.

68. Shapin and Schaffer, *Leviathan and the Air-Pump*.

69. Shapin and Schaffer, *Leviathan and the Air-Pump*, 60–65.

70. MCOA, Minutes of 1 Dec. 1868, 59.

71. MCOA, Minutes of 21 Jan. 1869, 71.

72. MCOA, Minutes of 13 April 1869, 17, 21–2.

73. MCOA, Minutes of 4 March 1869, 3–5.

74. Toussaint, *Dictionary of Mauritian Biography*, 448–9.

75. MCOA, Minutes of 17 June 1869, 27.

76. Barkly to Granville, undated, 49, RBGK-4.

77. Horne to Hooker, 26 Aug. 1869, Corr. 188.304, RBGK.

78. MCOA, Minutes of 22 March 1870, 3.

79. MCOA, Minutes of 19 July 1870, 60–61.

80. MCOA, Minutes of 13 Dec. 1870, 89; 16 Dec. 1870, 109.

81. Caldwell's Report, MCOA, Meeting of 11 Jan. 1871, 115–19.

82. MCOA, Minutes of 11 Jan. 1871, 114.

83. Hill, *Catalogue of Sugarcane*, 6, 34.

84. MCOA, Minutes of 9 Dec. 1869, 72.

85. MCOA, Minutes of 13 July 1869, 30; 26 July 1869, 38–9, 43; 28 Oct. 1869, 56–8.

86. MCOA, Minutes of 18 May 1871, 3.

87. RBGP, *Annual Report*, 1874.

88. RBGP, *Annual Report*, 1875, 4–5.

89. MCOA, Minutes of 21 Jan. 1869, 72; 4 March 1869, 6; 13 April 1869, 13–14; 28 Oct. 1869, 52–5; 7 Aug. 1884, 11–12; 20 April 1886, 22–6.

90. Carter, *Road to Botany Bay*, 1–33.

91. Public Record Office (PRO) CO 167/654 *Despatches 1890*.

92. MCOA, Minutes of 11 Jan. 1871, 107–8.

93. MCOA, Minutes of 9 Dec. 1869, 81.

94. MCOA, Minutes of 20 Jan. 1870, 91.

95. MCOA, Minutes of 15 March 1876, 52.

96. Drayton, "Imperial Science," 386–98, 404–7.

97. Drayton, "Imperial Science," 423–7, 435–6.

98. Horne to Hooker, Feb. 1870, Corr. 188.313, RBGK.

99. Horne to Hooker, 5 March 1874, Corr. 188.351, RBGK.

100. MCOA, Minutes of 23 March 1871, 143.

101. MCOA, Minutes of 30 Dec. 1874, 50.

102. MCOA, Minutes of 27 June 1884, 10.

103. MCOA, Minutes of 2 Feb. 1885, 23.

104. MCOA, Minutes of 11 Oct. 1875, 20.

105. RBGP, *Annual Report*, 1877, 3.

106. MCOA, Minutes of 22 Aug. 1878, 12–13.

107. MCOA, Minutes of 8 Aug. and 20 Sept. 1877.

108. *Revue agricole* 35, no.6 (1956): 324.

109. RBGP, *Annual Report*, 1879.

110. *Revue agricole* 35, no.6 (1956): 326.

111. MCOA, Minutes of 22 Aug. 1878, 13; 13 Nov. 1879, 44. Rouillard, *Historique de la canne à sucre*, 40.

112. RBGP, *Annual Report*, 1881, 16.

113. Horne to Dyer, 30 Nov. 1880, Corr. 188.398, RBGK.

114. MCOA, Minutes of 20 Sept. 1880, 14.

115. MCOA, Minutes of 5 July 1880, 10–12.

116. MCOA, Minutes of 28 March 1881, 26.

117. MCOA, Minutes of 16 April 1891, Annex, 3–4.

118. MCOA, Minutes of 16 April 1891, Annex, 3–4.

119. Horne to Dyer, 16 March 1881, Corr. 188.400, RBGK.

120. Horne to Dyer, 22 Nov. 1886, Corr. 188.430, RBGK.

121. Horne to Morris, 9 March 1887, Corr. 188.431, RBGK.

122. Horne to Hooker, 24 Jan. 1882, Corr. 188.408, RBGK. Horne to Dyer, 13 March 1882, Corr. 188.409, RBGK.

Notes to Chapter Three

1. Headrick, *Tentacles of Progress*, 215–16.

2. Rossiter, *Emergence of Agricultural Science.*

3. Fitzgerald, *Business of Breeding*, 12–22.

4. Latour, *The Pasteurization of France.*

5. Geison, *The Private Science of Louis Pasteur.*

6. Rossiter, *Emergence of Agricultural Science*, xi-xiv, 152–71.

7. Tinker, *New System of Slavery*, 242–3, 249–51, 255–8.

8. Grove, *Green Imperialism*, ch.5.

9. Mathur, "Partly-Elective Legislature," 76–7.

10. Mathur, "Partly-Elective Legislature," 77–8.

11. Mathur, "Partly-Elective Legislature," 84–96.

12. MCOA, Minutes of 12 April 1877, 3.

13. MCOA, Minutes of 17 May 1877, 14–15.

14. MCOA, Minutes of 22 Aug. 1878, 13–14.

15. Stevenson, *Genetics and Breeding*, 43.

16. Deerr, *Cane Sugar*, 32.

17. *Barbados Liberal*, 12 Feb. 1859.

18. G.K. Parris, "James W. Parris."

19. Fryer to Oliver, 23 June 1871, RBGK-8.

20. Stevenson, *Genetics and Breeding*, 45.

21. P. de Sornay, "Historique," 125–6. Rouillard, *Historique de la canne à sucre*, 21–2.

22. MCOA, Minutes of 11 Jan. 1871, 109–10.

23. P. de Sornay, "Historique," 125–6. Rouillard, *Historique de la canne à sucre*, 21–2.

24. Deerr, *Cane Sugar*, 42.

25. MCOA, Minutes of 3 May 1880, 7–8.

26. Colony of Mauritius, *Papers Relating to the Sugar Crisis.*

27. Summary of Newton's "The Sugar Crisis;" Pope Hennessy to Derby, 16 Feb. 1885; Dyer to Herbert 2 May 1885, 35–40, RBGK-1.

28. MCOA, Minutes of 5 April 1887, 17–21.

29. Stevenson, *Genetics and Breeding*, 46.

30. Stevenson, *Genetics and Breeding*, 47–8.

31. Carrington to Dyer, 11 May 1883, 1–2; 22 Nov. 1883, 14–15; Robinson to Dyer 28 Oct. 1883, 3; Robinson to Derby, 29 Oct. 1883, 5–6; "The Agricultural Society," *Barbados Agricultural Gazette*, Nov. 1887, 46; all in RBGK-10.

32. Wingfield to Dyer, 19 Nov. 1883, 7–11, RBGK-10.

33. *Report of the Results Obtained upon the Experimental Fields at Dodds Reformatory*, RBGK-10.

34. *Report of . . . Dodds Reformatory*, 1886–1887, RBGK-10.

35. "The Agricultural Society and the Dodds Experiments," *Barbados Agricultural Gazette*, Nov. 1887, 46, RBGK-10.

36. Harrison to Morris, 7 Jan. 1889, 47, RBGK-8.

37. *Report of . . . Dodds Reformatory*, 1889, RBGK-10. Demerara *Argosy*, 13 April 1889, 48, RBGK-8.

38. West India Committee, Circ. 41, 8 July 1889, 56, RBGK-8.

39. *Barbados Agricultural Gazette*, June 1891, 131–2, RBGK-10.

40. RBGK-10.

41. MCOA, Minutes of 11 Oct. 1886, 26, and 7 March 1889, 15. P. de Sornay, "Historique," 127.

42. Horne to Morris, 10 June 1890, Corr. 188.445, RBGK.

43. Horne to Dyer, 9 Dec 1890, Corr. 188.451, RBGK.

44. MCOA, Minutes of 2 May 1888, Annex, 1–2.

45. MCOA, Minutes of 2 May 1888, Annex, 3.

46. MCOA, Minutes of 24 Sept. 1888, 18.

47. *Report on the Proposed Creation of a Statn. Agr.*

48. MCOA, Minutes of 12 May 1890, 10.

49. MCOA, Minutes of 19 Feb. 1891, 8; 18 June 1891, sheet enclosed between 19–20.

50. MCOA, Minutes of 14 Dec. 1891, 33.

51. MCOA, Minutes of 18 April 1898, 6. P. de Sornay, "Historique," 129. Rouillard, *Historique de la canne à sucre*, 22. Robert, *Sugarcane Varieties*.

52. *Report on the Proposed Creation of a Statn. Agr.*

53. MCOA, Minutes of 16 April 1891, Annex, 1–2.

54. MCOA, Minutes of 19 Feb. 1891, 7–8.

55. MCOA, Minutes of 19 Feb. 1891, 7–8.

56. *Report on the Proposed Creation of a Statn. Agr.*

57. *Report on the Proposed Creation of a Statn. Agr.*, 15.

58. *Report on the Proposed Creation of a Statn. Agr.*, 15.

59. Shapin and Schaffer, *Leviathan and the Air-Pump*, 55.

60. *Report on the Proposed Creation of a Statn. Agr.*, 15.

61. Lees to Knutsford, with minutes by Wingfield, 9 Jan. 1891, PRO CO 167/661.

62. H. Leclézio, "Minutes of Proceedings of the Committee at a Meeting held at Government House, Port Louis, the 29th June 1892," enclosed with governor's despatch to the Colonial Office of 12 July 1892, PRO CO 167/669.

63. "Director of the Station Agronomique," 15–18 Aug. 1892 PRO 167/669 no.15464.

64. "Rules for the Working of the Station Agronomique," 168, RBGK-1. Minute by Wingfield, 4 Nov. 1892, PRO CO 167/670.

65. Station Agronomique, *Rapports et Bulletins*, 1900–1911.

66. MCOA, Minutes of 29 March, 10 May, 20 July, 1894.

67. Station Agronomique, *Rapports et Bulletins* 18, pp.14–15.

68. Rouillard, *Historique de la canne à sucre*, 22.

69. Rouillard, *Historique de la canne à sucre*, 19–20.

70. RBGP, *Annual Report*, 1891.

71. MCOA, Minutes of 12 Oct. 1894, 25; 3 April 1895, 18; 15 July 1895, 20; 2 Dec. 1895, 26; 15 Jan. 1895, 1–2; 7 April 1896, 3 and 2–3 of the Annex; 17 Nov. 1897, 30.

72. MCOA, Minutes of 7 April 1896, 3.

73. MCOA, Minutes of 28 Dec. 1898, 62; 15 Feb. 1900, 3; 24 Oct. 1901, 18.

74. Station Agronomique, *Rapports et Bulletins*, Annual Reports of 1900, 20–21; 1902, 24–6; 1904, 17–19; 1908, 49–51.

75. Station Agronomique, *Rapports et Bulletins* 15, 1–2.

76. Station Agronomique, *Rapports et Bulletins* 18, 1908, 14–15; 1909, 25–7.

77. Edgerton, *Sugarcane and Its Diseases*, 159–63. Ricaud, *Diseases of Sugarcane*, 358.

78. Stevenson, *Genetics and Breeding*, 48–9.

79. Stevenson, *Genetics and Breeding*, 52, 55, 56.

80. MCOA, Minutes of 11 April 1901, 8–9.

81. MCOA, Minutes of 7 April 1904, 20.

82. MCOA, Minutes of 29 July 1908, 2–3.

83. MCOA, Minutes of 26 Oct. 1900, 20.

84. Station Agronomique, Annual Report for 1909.

85. MCOA, Minutes of 3 Nov. 1904, 24–5; 29 Dec. 1905, 27; 7 March 1906, Annex; 25 April 1906, Annex; 7 Feb. 1907.

86. MCOA, Minutes of 8 Nov. 1907, Annex B; 30 Dec. 1908.

87. MCOA, Minutes of 7 April 1904, 19; 18 Aug. 1904, 2; 3 Nov. 1904, 24–5.

88. MCOA, Minutes of 3 Aug. 1911.

89. MCOA, Minutes of 11 April 1901, 8–9.

90. MCOA, Minutes of 11 April 1901, 9.

91. MCOA, Minutes of 3 July 1901, 12–13.

92. MCOA, Minutes of 17 Oct. 1902, 34; 27 Oct. 1902, 36; 27 April 1904, Committee Report Annex; 18 Aug. 1904, 2; 21 June 1905, 16; 21 June 1905, 17; 16 Aug. 1905, 21; 15 Nov. 1905, 25–6; 2 May 1906; 8 Aug. 1906, 16.

93. MCOA, Minutes of 1 July 1910, 2.

94. MCOA, Minutes of 18 July 1906, 11–12; 16 Aug. 1906, 19.

95. Latour, *Science in Action*, 108–21.

96. *Mauritius Chamber of Agriculture, 1853–1953.*

97. Lamusse, "Sources of Capital," 363–5.

98. Galloway, *Sugar Cane Industry*, 134.

99. Watts, *Mauritius Sugar Industry, 1929.*

100. Simmons, *Modern Mauritius*, 26–9.

101. Simmons, *Modern Mauritius.*

102. Horne to Dyer, 21 Dec. 1885, Corr. 188.425, RBGK.

103. RBGP, *Annual Reports*, 1891, 1893. Pike, *Rambles*, 126–7.

104. Hawley to Holland, 6 May 1887, 122–6, RBGK-1.

105. Horne, "Agricultural Resources of Mauritius, 1886;" *Annual Report of the Experimental Plantation Committee, 1888*, 5, RBGK-1.

106. Swettenham, *Royal Commission, 1909*, B:171–2.

107. Daniel Morris, "Memorandum of Observations" on Horne's "Agricultural Resources of Mauritius, 1886," 118, RBGK-1.

108. Allen, "Slender, Sweet Thread," 189–94.

109. Benedict, "Cash and Credit," 213–21; *Indians in a Plural Society*, 69–83.

110. Benedict, "Cash and Credit," 213–21. Ly-Tio-Fane Pineo, *Chinese Diaspora.*

111. Swettenham, *Royal Commission, 1909*, A:13–14, 19.

112. Swettenham, *Royal Commission, 1909*, A:10. *Report of the Small Planters' Commission, 1904–1905.*

113. Swettenham, *Royal Commission, 1909*, A:11.

114. MSIRI, *Annual Reports*, 1987, 12–14; 1988, 12–14; 1989, 14–19.

115. Reddi, "Political Awareness," 12–13.

116. Swettenham, *Royal Commission, 1909*, B:413–28.

117. Swettenham, *Royal Commission, 1909*, B:455.

Notes to Chapter Four

1. MCOA, Minutes of 18 July 1906, 13.

2. Drayton, "Imperial Science," 423–7, 435–6.

3. Golant, *Image of Empire. Bulletin of the Imperial Institute*, 1902–1948.

4. MCOA, Minutes of 16 Aug. 1906, 20; 9 April 1907, 15; 24 April 1907, 17; 26 June 1907, Annex; 15 July 1908, 1–3; 29 July 1908, 3; 4 Aug. 1909, 2.

5. "Sugar Cane Seed, from a Correspondent," Manchester *Examiner and Times*, 29 July 1890; Manchester *Courier*, 5 Aug. 1890; Morris, "Letter to the Editor," *European Mail*, 20 Aug. 1890; Harrison, "Letter to the Editor," Manchester *Examiner and Times*, 22 Aug. 1890; Harrison to Morris, 28 Aug. 1890, 72, RBGK-8.

6. Dyer to Secy. of State, 27 June 1893, RBGK-10.

7. *Agricultural Gazette*, Sept. 1892, 166, RBGK-10.

8. *West Indian Bulletin* 1, no.1 (1899): 4–5.

9. Morris to Dyer, 24 Dec. 1898, 18–19; 21 Jan. 1899, 20–21; 31 Jan. 1899, 57–62; 19 Jan. 1900; 30 March 1901, 102–3; 25 May 1901, 106–7; 6 Dec. 1901, 112–3; 16 Jan. 1902, 114–5; 28 Feb. 1902, 124–8; 1 Dec. 1902, 147; 3 Dec. 1907, 235, RBGK-11.

10. Jeffries, "Stockdale," 840–41.

11. Morris and Stockdale, "Improvement of the Sugar Cane," 345–50. Stevenson, *Genetics and Breeding*, 2.

12. Barnes, *Sugar Cane*, 25–7. Blackburn, *Sugar-Cane*, 38–41.

13. Morris and Stockdale, "Improvement of the Sugar Cane," 353–5.

14. *West Indian Bulletin* 1, no.1(1899): 13–14.

15. *West Indian Bulletin* 1, no.1(1899):16. Drayton, "Cane Breeding," 104. Morris to Dyer, 20 Dec. 1905, 209–10, RBGK-11. Morris to Dyer, 19 Jan. 1900, 57–62, RBGK-11.

16. For one example of the department's paternalism, see J.R. Williams, "Popular Agricultural Education in Jamaica," *West Indian Bulletin* 6 (1906): 227–33.

17. See comments by Morris in the *West Indian Bulletin* 11 (1911): 242.

18. Swettenham, *Royal Commission, 1909*, B:366–8.

19. MCOA, Minutes of 19 April 1911; 31 Jan. 1912.

20. Boyle to Harcourt, plus attached minutes, 23 Feb. 1911, PRO CO 167/796.

21. Minutes on the "Establishment of Dept. of Agriculture," June-July 1911, PRO CO 167/796 no.17400.

22. MCOA, Minutes of 2 Feb. 1912, Annex.

23. Harding to Collins, 25 April 1912, PRO CO 167/801.

24. Collins to Fiddes, Aug. 1912, 2, PRO CO 167/803 no.27762.

25. Collins to Prain, 11 May 1912; Prain to Collins, 27 May 1912; Harcourt to Chancellor, 19 June 1912, RBGK-2.

26. Chancellor to Harcourt, 3 Aug. 1912, 1, PRO CO 167/803.

27. Chancellor to Harcourt, 3 Aug. 1912, 6–7, PRO CO 167/803.

28. Chancellor to Harcourt, 3 Aug. 1912, 5–6, PRO CO 167/803.

29. Chancellor to Harcourt and to Fiddes, 3 Aug. 1912, PRO CO 167/803.

30. Collins to Fiddes, Aug. 1912, CO 167/803 no.27762, 2–3.

31. Harcourt to Prain, 3 Oct. 1912, RBGK-2.

32. MCOA, Minutes of 8 and 14 Nov. 1912.

33. Masefield, *Colonial Agricultural Service*, 161–2.

34. Jeffries, "Stockdale," 840.

35. Stockdale to Hill, 14 Feb. 1913, Corr. 188.553, RBGK.

36. MCOA, Minutes of 13 Feb. 1913.

37. Stockdale to Hill, 14 Feb. 1913, Corr. 188.553, RBGK.

38. Stockdale to Hill, 14 Feb. 1913, Corr. 188.553, RBGK.

39. A. North-Coombes, *Evolution of Sugarcane Culture*, 82–3.

40. Stockdale, "Preliminary Report," 26 May 1913, RBGK-2.

41. Stockdale to Hill, 7 June 1913, Corr. 188.554, RBGK.

42. *Revue Agricole*, no.43 (Jan.-Feb.1929): 1–3; no.54 (Nov.-Dec. 1930): 211–14.

43. *Revue Agricole* 42, no.2 (April-June 1963): 82–3.

44. *Revue Agricole*, 47, no.4 (Oct.-Dec. 1968): 254.

45. Manrakhan, *Mauritian School for Scientific Agriculture*, 17–29.

46. MCOA, President's Speech, 1914, 7.

47. MCOA, Circular of 20 July 1915.

48. MCOA, Minutes of 15 and 29 March 1916.

49. Stockdale to Hill, 10 June 1914, Corr. 188.556, RBGK.

50. Masefield, *Colonial Agricultural Service*, 63–9.

51. MCOA, Minutes of 20 April 1917.

52. Masefield, *Colonial Agricultural Service*, 162–3. *Revue agricole* 34, no.4 (July-Aug. 1955): 131. *Tropical Agriculture* 32, no.4 (Oct. 1955): 337–8.

53. Colony of Mauritius, Department of Agriculture, (COMDOA) *Annual Report*, 1914, 6.

54. COMDOA, *Annual Report*, 1918, 3.

55. COMDOA, *Annual Report*, 1917, 4.

56. Watts, *Mauritius Sugar Industry, 1929*.

57. Watts, *Mauritius Sugar Industry, 1929*, 15.

58. Worboys, "Science and the Colonial Empire," 19–21.

59. Masefield, *Colonial Agricultural Service*, 36–46, 63–74.

60. *Revue Agricole* 46, no.1 (Jan.-March 1967): 8–9.

61. *Revue Agricole* 46, no.1 (Jan.-March 1967): 10–11.

62. Manrakhan, *School for Scientific Agriculture*.

63. A. North-Coombes, interview with author. Floréal, Mauritius, 20 May 1992.

64. *Revue Agricole* 45, no.2 (April-June 1966): 58–9.

65. Manrakhan, *Mauritian School for Scientific Agriculture*, 320–21.

66. COMDOA, *Annual Report*, 1919, 3.

67. COMDOA, *Annual Reports*, 1922, 3; 1923, 3–4; 1924, 4–5; 1925, 3; 1926, 3. *Revue agricole*, no.7, Jan.-Feb. 1923, 48–50; no.15, May-June 1924, 181–4. A. North-Coombes, *Evolution of Sugarcane Culture*, 83–5.

68. COMDOA, *Annual Reports*, 1913, 9; 1914, 15; 1924, 12; 1929, 8.

69. COMDOA, *Annual Report*, 1917, 7.

70. COMDOA, *Annual Report*, 1917, 2.

71. COMDOA, *Annual Report*, 1938, 11.

72. COMDOA, *Annual Report*, 1919, 3.

73. Jepson, *Phytalus Investigation, 1936*.

74. Rouillard, *Historique de la canne à sucre*, 41.

75. COMDOA, *Annual Report*, 1919, 4. Tempany, *Memorandum on Experiments with Varieties of Sugar Cane*, 9 April 1919, 1–2.

76. COMDOA, *Annual Report*, 1918, 17; 1928, 17.

77. Rouillard, *Historique de la canne à sucre*, 27, 29.

78. *Revue agricole* 19, 1 (1940): 8. Stevenson, *Genetics and Breeding*, 51.

79. *Revue agricole* no. 40 (1928): 187–90.

80. *Revue agricole* no. 37, Jan.-Feb. 1928, 23–5.

81. *Revue agricole* no. 37, Jan.-Feb. 1928, 12, 30.

82. COMDOA, *Annual Report*, 1928, 17.

83. COMDOA, Sugarcane Research Station (SRS), *Annual Reports*, 1934, 5; 1935, 1; 1936, 5.

84. Barber, *Tropical Agricultural Research*, 10–11.

Notes to Chapter Five

1. Chancellor to Secretary of State, 22 May 1911. *Further Despatches Relative to the Mauritius Royal Commission of 1909*, 22.

2. Wilberforce, *Cooperative Credit Banks*, 6.

3. Wilberforce, *Cooperative Credit Banks*, 8–9.

4. Wilberforce, *Cooperative Credit Banks*, 7.

5. COMDOA, *Cooperative Credit Societies*, 1914, 4; 1915, 1; 1917, 2; 1926, 1–2; 1935, 9; 1937, 10.

6. M.D. North-Coombes, "Struggles," 27.

7. Benedict, "Cash and Credit."

8. COMDOA, *Cooperative Credit Societies*, 1914, 3.

9. COMDOA, *Cooperative Credit Societies*, 1924, 2; 1926, 3.

10. Colony of Mauritius, *Census* of 1911, 1921, and 1931.

11. COMDOA, *Annual Reports*, 1916, 15; 1923, 13.

12. Colony of Mauritius, *Land Settlement Report*, 1944.

13. COMDOA, *Annual Reports*, 1913–1937. *Revue agricole*, no.12, Nov.-Dec. 1923, 331. COMDOA, Bulletin No.26, "Hints on the Establishment and Management of School Gardens," Jan. 1922.

14. Stockdale to Col. Secy., 26 Jan. and 7 Feb. 1914, 145–50, RBGK-2. Stockdale to Hill, 10 July 1914, Corr. 188.558, RBGK.

15. Green and Hymer, "Cocoa in the Gold Coast," 299, 306–11.

16. F. North-Coombes, *Revue agricole* no. 1, Jan.-Feb. 1922, 1–2.

17. Watts, *Mauritius Sugar Industry, 1929.*

18. "Visit of Sir Francis Watts to Mauritius" and minute by Stockdale, 4 April 1929, PRO CO 167/864/4 no.64589.

19. A. North-Coombes, interview with author. Floréal, Mauritius, 20 May 1992.

20. Simmons, *Modern Mauritius*, 29–34, 46–7.

21. Simmons, *Modern Mauritius*, 54. *Mauritius Chamber of Agriculture, 1853–1953.*

22. Simmons, *Modern Mauritius*, 58–63.

23. M.D. North-Coombes, "Struggles," 20–22.

24. *Revue agricole* 35, no.1 (Jan.-Feb. 1956): 7.

25. *Revue agricole* 13, no.78 (1934): 208.

26. COMDOA, SRS, *Annual Report*, 1937, 37.

27. COMDOA, *Annual Report*, 1931, 3–4.

28. *Revue agricole* no. 61 (1932): 6.

29. Rouillard, *Historique de la canne à sucre*, 28, 30.

30. COMDOA, *Annual Report*, 1932, 13–14.

31. Fitzgerald, *Business of Breeding*, 23–30. Griffing, "Contributions of Quantitative Genetics," 44–50. Stebbins, "Biological Revolutions," 5–6.

32. Stevenson, *Genetics and Breeding*, 3.

33. Stevenson, *Genetics and Breeding*, 49–50.

34. Jeswiet, "The History of Sugar-Cane Selection Work."

35. Albert and Graves, *World Sugar Economy*, 15.

36. Kohler, *Lords of the Fly*, 11–13, 133–7.

37. Starr and Taggart, *Biology*, 177–83.

38. Simmonds, *Evolution of Crop Plants*, 106–7.

39. Stevenson, *Genetics and Breeding*, chs.4 and 6. Venkatraman, "Methods of Selecting Seedlings" and "Problems for the Sugar Cane Breeder."

40. COMDOA, *Annual Report*, 1932, 21.

41. *Revue agricole* no.34 (July-Aug. 1927): 244–5; v.38, no.5 (Sept.-Oct. 1959): 213.

42. *Revue agricole* 9, no.52 (1930): 147.

43. *Revue agricole* 10, no.60 (1931):220–32; 13, no.73 (1934):30; 13, no.81 (1935):106; 13, no.85 (1936):12.

44. Aimé de Sornay, "Méthode d'obtention de nouvelles variétés de cannes."

45. COMDOA, *Annual Report*, 1931, 3–4.

46. Aimé de Sornay, "La canne à sucre M.134/32," 6–7, 11.

47. COMDOA, *Annual Report*, 1936, 15–16, 21; COMDOA, SRS, *Annual Reports*, 1935–1936; *Revue agricole* no.81 (1935): 97; no.85 (1936): 13.

48. MCOA, President's Report of 1867, 1.

49. Stevenson, *Genetics and Breeding*, 27–8.

50. MCOA, Minutes of 7 Dec. 1898, 61; 28 Dec. 1899.

51. MCOA, Minutes of 7 April 1904, 20.

52. COMDOA, *Annual Report*, 1914, 3.

53. Stevenson, *Genetics and Breeding*, 28.

54. Haddom, "Milling of Uba," 13–14.

55. Stevenson, *Genetics and Breeding*, 29.

56. COMDOA, *Annual Reports*, 1927, 16; 1928, 9; 1933, 16–17; SRS, *Annual Report*, 1933, 35. Stevenson, "Origin of Uba," 4–7.

57. Hooper, *Commission of Enquiry, 1937*, 135.

58. Hooper, *Commission of Enquiry, 1937*, 133–5.

59. *Revue agricole* no.72 (1933): 192–5.

60. F. North-Coombes, *Mes champs et mon moulin*, 355–6.

61. Hooper, *Commission of Enquiry, 1937*, 135–7.

62. Hooper, *Commission of Enquiry, 1937*, 9–10.

63. Hooper, *Commission of Enquiry, 1937*, 9.

64. Hooper, *Commission of Enquiry, 1937*, 10–11.

65. Hooper, *Commission of Enquiry, 1937*, 11–14.

66. Hooper, *Commission of Enquiry, 1937*, 14–15.

67. Hooper, *Commission of Enquiry, 1937*, 15–16.

68. Hooper, *Commission of Enquiry, 1937*, 16–18.

69. Hooper, *Commission of Enquiry, 1937*, 16–18.

70. Hooper, *Commission of Enquiry, 1937*, 18–22.

71. Hooper, *Commission of Enquiry, 1937*, 22–3.

72. Hooper, *Commission of Enquiry, 1937*, 27–9, 212.

73. Hazareesingh, *History of Indians in Mauritius*.

74. M.D. North-Coombes, "Struggles," 37–8.

75. Hooper, *Commission of Enquiry, 1937*, 133–4.

76. Hooper, *Commission of Enquiry, 1937*, 143.

77. Bede to Dufferin, "Constitution, 1938," PRO CO 167/900/9 no.57249.

78. Hooper, *Commission of Enquiry, 1937*, 133–4. COMDOA, *Annual Report*, 1937, 15–17. Hill, *Catalogue*, 10.

79. COMDOA, *Annual Report*, 1938, 19; SRS, *Annual Reports*, 1938, 5–6, 38–40; 1939, 5, 25; 1940, 15–16; 1942, 14–15; 1943, 15; 1944, 23.

80. Craig, "Uba Replacement," 52–8.

81. COMDOA, *Annual Reports*, 1951, 27; 1954, 58; SRS, *Annual Report*, 1948, 44.

82. *Revue agricole* 25, no.3 (1946): 102; no.5 (1946): 216–7.

83. Shapin and Schaffer, *Leviathan and the Air-Pump*, 332.

84. Headrick, *Tentacles of Progress*, 13.

Notes to Chapter Six

1. Morgan, *Colonial Development*, 1:14–15.

2. Simmons, *Modern Mauritius*, 73–7.

3. Rouillard, *Historique de la canne à sucre*, 30.

4. *The Mauritius Chamber of Agriculture, 1853–1953*, and MCOA, *President's Report*, 1953.

5. *Revue agricole* 20, no.3 (1941): 138.

6. *Revue agricole* 16, no.105 (1939): 71.

7. *Revue agricole* 22, no.1 (1943): 22.

8. Rouillard, *Historique de la canne à sucre*, 32.

9. Masefield, *Colonial Agricultural Service*, 44–5.

10. COMDOA, *Annual Report*, 1947, 2–3.

11. COMDOA, *Annual Report*, 1951, 5.

12. *Revue agricole* 20, no.4 (1941): 212.

13. *Revue agricole* 26, no.3 (1947): 151, 205–6; 27, no.3 (1948): 135.

14. *Revue agricole* no.73 (1934):7–9; no.79 (1935):54–62.

15. Craig, "Cane Varieties from Coimbatore," 4–8.

16. *Revue agricole* no.91 (1937): 24, 30; no.97 (1938): 20–22.

17. COMDOA, SRS, *Annual Report*, 1935, 1.

18. A. North-Coombes, interview with author. Floréal, Mauritius, 20 May 1992.

19. Internal records of the Chamber of Agriculture for the period after 1940 are not yet public, but I was able to locate some confidential sources.

20. COMDOA, *Annual Report*, 1946, 2. *Revue agricole* 28, no.5 (1949): 247.

21. *Revue agricole* 28, no.2 (1949): 55–61.

22. Rouillard, *Historique de la canne à sucre*, 31–2.

23. COMDOA, *Annual Report*, 1949, 1–2.

24. *Revue agricole* 26, no.3 (1947): 207.

25. Simmons, *Modern Mauritius*, 78–90.

26. Bowman, *Mauritius*, 44.

27. Simmons, *Modern Mauritius*, 96–102.

28. Simmons, *Modern Mauritius*, 123–4.

29. *Mauritius Economic Commission 1947–48*, 2:35–44. *Revue agricole* 27, no.5 (1948):236; 27, no.4 (1948): 191–2.

30. *Revue agricole* 28, no.5 (1949): 245.

31. *Revue agricole* 29, no.1 (1950): 9–10.

32. *Revue agricole* 30, no.4 (1950): 190.

33. *Revue agricole* 27, no.3 (May-June 1948): 94–5; 36, no.2 (March-April 1957): 61–2.

34. Colony of Mauritius, *Legislative Council Debates*, 23 June 1953, 31; 26 June 1953, 14–15; 30 June 1953, 7.

35. Simmons, *Modern Mauritius*, 115.

36. Simmons, *Modern Mauritius*, 128–34.

37. Bowman, *Mauritius*, 104.

38. Koenig, *Mauritius and Sugar*.

39. Koenig, "Commonwealth Sugar Agreement."

40. Maier, *Recasting Bourgeois Europe*.

41. Eriksen, *Communicating Cultural Difference*.

42. Bowman, *Mauritius*, 69–82.

43. Bowman, *Mauritius*, 82–100.

44. Houbert, "Mauritius," 90–91.

45. Bowman, *Mauritius*, ch.5.

46. MCOA, *The Sugar Industry: Situation and Outlook*.

47. Jacques Koenig. Interview with author, Port Louis, Mauritius, 11 Sept. 1992.

48. MSIRI, *Annual Report*, 1966, 10.

49. MSIRI, *Annual Report*, 1968, 10.

50. MSIRI, *Annual Report*, 1969, 9–10.

51. MSIRI, *Annual Report*, 1972, 8.

52. MSIRI, *Annual Report*, 1987, 3; 1988, 3; 1989, 5; 1990, 7.

53. Bartholomew, "Modern Science in Japan."

54. Meade, *Economic and Social Structure*, 77–8. World Bank, *Staff*

Appraisal Report No.6092–MAS, 5 June 1985, 9.

55. Domaingue, "Canne à sucre," 6–7. MCOA, *Annual Report*, 1988–1989, 53.

56. A. de Sornay, "Resistance," 241–5.

57. Tyack, "Pensées vertes," 13–14.

58. MCOA, *Planters' Sugar Cane Yields*, 1971; *Planters' Yields*, 1975; *Planters' Sugar Cane Yields, Second Report*, 1980.

59. COMDOA, *Annual Reports*, 1947, 29; 1948, 3, 16–17; 1949, 28; 1950, 5, 27; 1951, 5, 26–7; 1952, 5; 1953, 5.

60. Pillay, "Redesigning Extension Services," 78.

61. MCOA, *Planters' Sugar Cane Yields*, 1971, 15.

62. Lutchmeenaraidoo, *Committee of Enquiry*, 43.

63. MCOA, *Planters' Sugar Cane Yields*, 1971, 17; *Planters' Yields*, 1975, 2–3.

64. MCOA, *Planters' Sugar Cane Yields*, 1980, 61.

65. Pillay, "Redesigning." Banker, "Problems of Small Planters." "Interview with Bruce Campbell," *PROSI*, 1988.

Notes to the Conclusion

1. Eriksen, "Nationalism, Mauritian Style," 554–5.

2. Gilroy, *Black Atlantic*.

3. Hookoomsing, "So Near," 33.

4. Bowman, *Mauritius*, 54.

5. Shapin and Schaffer, *Leviathan and the Air-Pump*.

6. Berry, *No Condition Is Permanent*, ch.2.

7. Chambers, *Rural Development*, 92–101.

8. Lipton, *New Seeds and Poor People*, 28–31.

BIBLIOGRAPHY

I. Primary Sources

Archives of the Royal Botanic Gardens, Kew (RBGK).
Correspondence of James Duncan, J.B. Harrison, John Horne, Daniel Morris, Nicolas Pike, William Scott, and Frank Stockdale.
Miscellaneous Reports (MR). (RBGK-1–11 abbreviations are my own.)
RBGK-1 MR 11.4: *Mauritius. Agriculture. 1885–1900.*
RBGK-2 MR 11.4: *Mauritius. Agriculture. Rodriguez. Forests, &c.*
RBGK-3 MR 11.4: *Mauritius. Botanic Garden. 1860–1919.*
RBGK-4 MR 11.4: *Mauritius. Botanic Garden. 1865–1893.*
RBGK-5 MR 15.6: *Barbados. Agricultural Work, 1899–1907.*
RBGK-6 MR 15.6: *Barbados. Agriculture. Agricultural Products, &c. 1884–1905.*
RBGK-7 MR 15.6: *Barbados. Commissioner of Agriculture. 1898–1909.*
RBGK-8 MR 15.6: *Barbados. Sugar Cane Cultivation, Dept. of Agriculture, &c. 1871–1918.*
RBGK-9 MR 15.6: *Barbados. Sugar Cane Diseases. 1889–1901.*
RBGK-10 MR 15.6: *Barbados. Sugar Cane Experiments. 1883–1900.*
RBGK-11 MR 15.6: *West Indies. Commissioner of Agriculture Letters, 1898–1914.*

Public Record Office, Kew (PRO). Colonial Office records (CO).
Despatches and Minutes, Mauritius, 1890–92, 1911–13: CO 167/654, 661, 669, 670, 796, 801, 803, 807.
Sugar Industry Research. 1928. CO 167/861/5 no.54608.
Agricultural Research Station, Proposed Establishment of, 1928. CO 167/863/12 no.64569.
Visit of Sir Francis Watts to Mauritius. 1929. CO 167/864/4 no.64859.
Constitution. 1937. CO 167/897/15 no.57249.
Constitution. 1938. CO 167/900/9 no.57249

Mauritius Chamber of Agriculture Library, Port Louis (MCOA).
The Mauritius Chamber of Agriculture. *The Sugar Industry: Situation and Outlook: Elements for a Strategic Reappraisal.* November 1991.

211

Minutes of Meetings, 1868 to 1940.
President's Reports, 1853 to the Present.

Mauritius Archives and the University of Mauritius Library.

Avramovic, Dragoslav. *Report of the Commission of Enquiry on the Sugar Industry.* Port Louis: Government Printer, 1984.

Balogh, T., and C.J.M. Bennett. *Commission of Enquiry: Sugar Industry 1962.* Port Louis: Government Printer, 1963.

Berthelot, T.B., and K.P. Pillay. *Structure of Agricultural Research and Linkage with Extension: Mauritian Agriculture at the Crossroads.* MSIRI, 1988.

Bett, J. *Report on the Possibilities of Land Settlement in Mauritius.* Development and Welfare Department, 1948.

Burrenchobay, M. *Catechism of Co-Operative Credit.* Department of Agriculture, Bulletin No.46, General Series. Port Louis: Government Printer, 1937.

——. *Guide for Use by Committees of Cooperative Credit Societies of Unlimited Liability Type.* Department of Cooperation, 1950.

——. *A Guide to the Operation of the Cooperative Credit Societies in Mauritius.* Department of Agriculture, 1936

——. *A Short Account of the Cooperative Credit Societies Movement in Mauritius and the Legislation by Which It Is Controlled.* Department of Agriculture, 1944.

Campbell, W.K.H. *Cooperation: A Series of Talks Given on the Mauritius Broadcasting Service.* Port Louis: Government Printing Office, March 1945.

——. *Cooperation in Mauritius.* Port Louis: Government Printing Office, 1945.

Colony of Mauritius. Department of Agriculture. *Annual Reports,* 1914–1957.

Colony of Mauritius. Department of Agriculture. *Reports on the Working of Cooperative Credit Societies in Mauritius,* 1914–1947.

Colony of Mauritius. Department of Agriculture. *Regulations and Syllabus of the School of Agriculture.* Bulletin No.10, General Series. June 1918.

Colony of Mauritius. Department of Agriculture. Sugarcane Research Station. *Annual Reports,* 1930–1953.

Colony of Mauritius. Department of Cooperation. *Annual Reports,* 1948–1959.

Colony of Mauritius. *Legislative Council Debates.*

Colony of Mauritius. *Mauritius Land Settlement Report.* 1944.

Colony of Mauritius. *Papers Relating to the Sugar Crisis.* 1885.

Colony of Mauritius. Royal Botanic Gardens, Pamplemousses. *Annual Reports,* 1865–1893.

Craig, N. *Sugarcane in Mauritius: A Revue of the Present Position with Respect to its Cultivation and Manuring.* Sugarcane Research Station Bulletin No.11. Port Louis: Government Printer, 1937.

Duncan, James. *Catalogue of Plants in the Royal Botanical Garden, Mauritius.* Port Louis: H. Plaideau, 1863.

Dupont, P.R. *Hints on the Establishment and Management of School Gardens.* Department of Agriculture, Bulletin No.26, General Series. Port Louis: Government Printer, 1922.

Further Despatches Relative to the Recommendations of the Mauritius Royal Commission of 1909. Presented to the Council of Govt by the Governor, June 20, 1911.

Hart, W. Edward. *Le jardin botanique des Pamplemousses: notice historique* (The Pamplemousses botanic garden). Port Louis: Imprimerie du Gouvernement, 1916.

Hill, A. Glendon. *An Annotated Catalogue of Sugarcane Varieties in Mauritius, Present and Past.* Sugarcane Research Station, Bulletin No.13. Port Louis: Government Printer, 1937.

Hooper, Charles Arthur, et al., *Colony of Mauritius. Report of the Commission of Enquiry into Unrest on Sugar Estates in Mauritius, 1937.* Port Louis: R.W. Brooks, Government Printer, 1938.

Horne, John. *Report on the Agricultural Resources of Mauritius.* 1886.

The Interim Report of the Small Cultures Committee, Laid before the Council of Government. July 27, 1909.

Jepson, W.F. *A Summary of the Results of the Phytalus Investigation, 1933–1936, with Recommendations as to Further Lines of Work; being a General Report on Studies Carried out on behalf of the Mauritius Sugar Industry Reserve Fund.* 1936.

Koenig, Maxime. *Agricultural Census in Mauritius.* 1930.

Lutchmeenaraidoo, K., et al. *Committee of Enquiry: Sugar Industry: 1972–1973.* Port Louis: Government Printer, 1973.

Manrakhan, J., and R. Sithanen, *Report of the Commission of Enquiry on the Sugar Industry: The Human Factor, the Sugar Industry, and Mauritius: A Vision for the Future.* Port Louis: Government Printer, 1984.

Mauritius Sugar Industry Research Institute. *Annual Reports,* 1954–1991.

Ministry of Agriculture and Natural Resources. *Technical Report. Survey of Sugar Cane Planters and Their Production Pattern.* March 1973.

Moody, S., et al. *Report of the Commission of Enquiry into the Disturbances Which Occurred in the North of Mauritius in 1943.* Port Louis: Government Printer, 1944.

O'Connor, C.A. *Report on Tour in the Union of South Africa.* 1930.

Robert, Henri. *Sugarcane varieties in Mauritius.* Department of Agriculture, Bulletin No. 2, Statistical Series. Port Louis: Government Printer, 1915.

Report of the Mauritius Economic Commission, 1947–1948. 2 Parts. Port Louis: Government Printers, 1948.

Report of the Small Planters' Commission, with Proceedings and Evidence, 1904–1905. Port Louis.

Report on the Proposed Creation of a Station Agronomique, Adopted in Committee on 2nd December 1890.

Saxena, M.P. *Mauritius Summary Report on Survey of Agricultural Cooperative Credit Societies in Mauritius, 1973–74.* December 1976.

Shepard, E.F.S. *Report on Visit to Australia and Java.* 1931.

Station Agronomique. *Rapports et Bulletins. Deuxième Série* (Reports and bulletins. Second series). 1900–1911.

Stevenson, G.C. *An Investigation into the Origin of the Sugarcane Variety Uba Marot.* Sugarcane Research Station, Bulletin No.17. Port Louis: Government Printer, 1940.

———. *Sugarcane Varieties Produced by the Sugarcane Research Station, and Their Value to the Sugar Industry of Mauritius.* Sugarcane Research Station, Bulletin No.18. Port Louis: Government Printer, 1946.

Swettenham, Frank, Edward O'Malley and Hubert Woodcock. *Report of the Mauritius Royal Commission, 1909.* London: HMSO, 1910.

Tempany, H.A. *Memorandum on Experiments with Varieties of Sugar Cane,* 9 April 1919, Department of Agriculture.

Watts, Francis. *Report on the Mauritius Sugar Industry, 1929.* London: HMSO, 1930.

Wilberforce, S. *Report on the Prospects of Cooperative Credit Banks among Indian Small Planters in Mauritius.* July 2, 1913.

Mann Library of Cornell University and the U.S. National Agricultural Library.

Barbados. Department of Agriculture. *Report on the Sugar-Cane Experiments,* 1919–1926.

British Guiana. Department of Agriculture. *Sugar Bulletin,* 1–23, 1933–1953.

Bulletin of the Imperial Institute: A Quarterly Record of Progress in Tropical Agriculture and Industries and the Commercial Utilisation of the Natural Resources of the Dominions, Colonies, and India. London: John Murray, 1902–1948.

Department of Science and Agriculture, Barbados. *Agricultural Journal* 1–8, 1932–1939.

Report of the Department of Agriculture, Barbados, 1910–1931.

West Indian Bulletin: The Journal of the Imperial Department of Agriculture for the West Indies, 1–18, 1900–1921.

II. Secondary sources

Adas, Michael. *Machines as the Measure of Men: Science, Technology, and Ideologies of Western Dominance.* Ithaca: Cornell University Press, 1988.

Albert, Bill, and Adrian Graves, eds. *The World Sugar Economy in War and Depression, 1914–1940.* London: Routledge, 1988.

Allen, Richard Blair. "Creoles, Indian Immigrants, and the Restructuring of Society and Economy in Mauritius, 1767–1885." Ph.D. diss., Department of History, University of Illinois at Urbana-Champaign, 1983.

——. "The Slender, Sweet Thread: Sugar, Capital, and Dependency in Mauritius, 1860–1936." *Journal of Imperial and Commonwealth History* 16, no.2 (1988): 176–200.

Alpers, Edward A., "The French Slave Trade in East Africa, 1721–1810." *Cahiers d'études africaines* 37 (1970): 80–124.

Ambrosoli, Mauro. *The Wild and the Sown: Botany and Agriculture in Western Europe, 1350–1850.* Cambridge: Cambridge University Press, 1997.

Arno, Toni and Claude Orian. *Ile Maurice: une société multiraciale* (Mauritius: a multiracial society). Paris: Editions l'Harmattan, 1986.

Baber, Zaheer. *The Science of Empire: Scientific Knowledge, Civilization, and Colonial Rule in India.* Albany: State University of New York Press, 1996.

Baker, Philip. *Kreol: A Description of Mauritian Creole.* London: Hurst, 1972.

Baker, Philip, and C. Corne. *Isle de France Creole: Affinities and Origins.* Ann Arbor, Michigan: Karoma, 1982.

Baker, Philip, and Vinesh Y. Hookoomsing. *Diksyoner kreol morisyen* (Dictionary of Mauritian Kreol). Paris: Editions l'Harmattan, 1987.

Banker, T. "An Analysis of the Problems of Small Planters in Mauritius, with Special Attention to the Farmers' Service Centres." Diploma Essay in Development Studies, University of Mauritius, 1990.

Barber, C.A. *Tropical Agricultural Research in the Empire, with Special Reference to Cacao, Sugar Cane, Cotton, and Palms.* Empire Marketing Board Publication No. 2. London: HMSO, 1927.

Barnes, A.C. *The Sugar Cane.* New York: John Wiley, 1974.

Bartholomew, James R. "Modern Science in Japan: Comparative Perspectives." *Journal of World History* 4, no.1 (1993): 101–116.

Basalla, George. "The Spread of Western Science," *Science* 156 (5 May 1967): 611–622.

Bayly, C.A. *Imperial Meridian: The British Empire and the World, 1780–1830.* London: Longman, 1989.

Beckles, Hilary M. *A History of Barbados: From Amerindian Settlement to Nation-State.* Cambridge: Cambridge University Press, 1990.

Benedict, Burton. "Cash and Credit in Mauritius." *Journal of South African Economics* 26, no.3 (1958): 213–221.

——. *Indians in a Plural Society: A Report on Mauritius.* Colonial Office. Colonial Research Study No. 34. London: H.M.S.O., 1961.

——. *Mauritius: The Problems of a Plural Society.* London: Pall Mall Press, 1965.

Berman, Bruce J. *Control and Crisis in Colonial Kenya: The Dialectic of Domination.* Athens, Ohio: Ohio University Press, 1990.

Berry, Sara S. *Fathers Work for Their Sons: Accumulation, Mobility, and Class Forma-tion in an Extended Yorùbá Community*. Berkeley and Los Angeles: University of California Press, 1985.

————. *No Condition Is Permanent: The Social Dynamics of Agrarian Change in Sub-Saharan Africa*. Madison: University of Wisconsin Press, 1993.

Biagioli, Mario. *Galileo, Courtier: The Practice of Science in the Culture of Absolut-ism*. Chicago: University of Chicago Press, 1993.

Bijker, Wiebe E., Thomas P. Hughes, and Trevor J. Pinch, eds. *The Social Construc-tion of Technological Systems: New Directions in the Sociology and History of Technology*. Cambridge: MIT Press, 1987.

Bissoondoyal, U., ed. *Indian Labour Immigration*. Moka, Mauritius: Mahatma Gandhi Institute, 1986.

————, ed. *Indians Overseas: The Mauritian Experience*. Moka, Mauritius: Ma-hatma Gandhi Institute, 1984.

————, ed. *Slavery in South-West Indian Ocean*. Moka, Mauritius: Mahatma Gandhi Institute, 1989.

Blackburn, Frank. *Sugar-Cane*. New York: Longman, 1984.

Blume, Helmut. *Geography of Sugar Cane: Environmental, Structural, and Economi-cal Aspects of Cane Sugar Production*. Berlin: Verlag Dr. Albert Bartens, 1985.

Bojer, Wenzel. *Hortus Mauritianus: ou énumeration des plantes exotiques et indigènes, qui croissent à l'île Maurice, disposées d'apres la methode naturelle* (The Mauritian garden: or enumeration of the exotic and indigenous plants that grow in Mauritius, arranged after the natural method). Port Louis: Aimé Mamarot et Cie., 1837.

Bonneuil, Christophe. *Des savants pour l'empire: la structuration des recherches scientifiques coloniales au temps de "la mise en valeur des colonies françaises," 1917–1945*. (Wise men for the empire: the structuring of colonial scientific research during the "development" era of the French colonies). Paris: Edi-tions de l'ORSTOM, 1991.

Boullé, Bernard, ed. *The Sugar Protocol and Mauritius*. Port Louis: Editions IPC, 1989.

Bouton, Louis. *Rapport présenté à la Chambre d'Agriculture sur les diverses éspeces de canne à sucre cultivées à Maurice* (Report presented to the Chamber of Agri-culture on the different kinds of sugar canes cultivated in Mauritius). Port Louis: Channel, 1863.

Bowman, Larry W. *Mauritius: Democracy and Development in the Indian Ocean*. Boulder, Colorado: Westview Press, 1991.

Boyd, Mark F., ed. *Malariology: A Comprehensive Survey of All Aspects of This Group of Diseases from a Global Standpoint*. 2 vols. Philadelphia and London: W.B. Saunders Company, 1949.

Brandes, E.W., and G.B. Sartoris. "Sugarcane: Its Origin and Improvement." In

United States Department of Agriculture. *Yearbook of Agriculture: 1936.* Washington, D.C.: USGPO, 1936, 561–611.

Brockway, Lucile H. *Science and Colonial Expansion: The Role of the British Royal Botanic Gardens.* New York: Academic Press, 1979.

Burroughs, Peter. "The Mauritius Rebellion of 1832 and the Abolition of British Colonial Slavery." *Journal of Imperial and Commonwealth History* 4, no.3 (1976): 243–265.

Busch, Lawrence, ed. *Science and Agricultural Development.* Totowa, New Jersey: Allenheld & Osmun, 1981.

Busch, Lawrence, and William B. Lacy. *Science, Agriculture, and the Politics of Research.* Boulder, Colorado: Westview Press, 1983.

Busch, Lawrence, William B. Lacy, Jeffrey Burkhardt, and Laura R. Lacy. *Plants, Power, and Profit: Social, Economic, and Ethical Consequences of the New Biotechnologies.* Oxford: Basil Blackwell, 1991.

Cain, P.J., and A.G. Hopkins. *British Imperialism.* Volume One: *Innovation and Expansion, 1688–1914.* Volume Two: *Crisis and Deconstruction, 1914–1990.* London: Longman, 1993.

Carter, Marina. *Lakshmi's Legacy: The Testimonies of Indian Women in 19th Century Mauritius.* Rose Hill: Éditions de l'Océan Indien, 1994.

———. "Strategies of Labour Mobilisation in Colonial India: The Recruitment of Indentured Workers for Mauritius." *Journal of Peasant Studies* 19, nos.3–4 (1992): 229–245.

———. *Voices from Indenture: Experiences of Indian Migrants in the British Empire.* London: Leicester University Press, 1996.

Carter, Marina, and Hubert Gerbeau. "Covert Slaves and Coveted Coolies in the Early Nineteenth-Century Mascareignes." *The Economics of the Indian Ocean Slave Trade in the Nineteenth Century.* Ed. W.G. Clarence-Smith. Totowa, New Jersey: Frank Cass, 1989, 194–208.

Carter, Paul. *The Road to Botany Bay: An Exploration of Landscape and History.* New York: Alfred A. Knopf, 1987.

Centenaire de la Société Royale des Arts et des Sciences de l'île Maurice, 1829–1929 (Centenary of the Royal Society of Arts and Sciences of Mauritius). Port Louis: Typographie Moderne, 1932.

Chambers, Robert. *Rural Development: Putting the Last First.* Harlow, Essex: Longman, 1983.

Chan Low, L. Jocelyn. "T'Eylandt Mauritius: la précolonisation: 1598–1638" (The island of Mauritius before colonization). *Journal of Mauritian Studies* 4, no.1 (1992): 36–65.

Chaudenson, Robert. "La situation linguistique dans les archipels créolophones de l'Océan Indien" (The linguistic situation in the creole-speaking islands of the Indian Ocean). *Annuaire des pays de l'Océan Indien.* (1974): 154–182.

Cohen, David William. *The Combing of History*. Chicago: University of Chicago Press, 1994.

Cohen, David William, and E.S. Atieno Odhiambo. *Siaya: The Historical Anthropology of an African Landscape*. Athens: Ohio University Press, 1989.

Comaroff, Jean, and John Comaroff. *Of Revelation and Revolution: Christianity, Colonialism, and Consciousness in South Africa*. Vol. 1. Chicago: University of Chicago Press, 1991.

Craig, N. "The Introduction to Mauritius of Cane Varieties Produced by the Experimental Station of Coimbatore." *Revue agricole et sucrière de l'île Maurice* 19, no.1 (1940).

———. "The Uba Replacement Scheme." *Revue agricole et sucrière de l'île Maurice* 23, no.2 (1944): 52–58.

Curtin, Philip D. *Cross-Cultural Trade in World History*. Cambridge: Cambridge University Press, 1984.

———. *Death by Migration: Europe's Encounter with the Tropical World in the Nineteenth Century*. Cambridge: Cambridge University Press, 1989.

———. *The Rise and Fall of the Plantation Complex: Essays in Atlantic History*. Cambridge: Cambridge University Press, 1990.

Dean, Warren. "The Green Wave of Coffee: Beginnings of Tropical Agricultural Research in Brazil, 1885–1900." *Hispanic American Historical Review* 69, no.1 (1989): 91–115.

Dear, Peter. "Totius in Verba: Rhetoric and Authority in the Early Royal Society." *Isis* 76 (1985): 145–161.

Deerr, Noël. *Cane Sugar: A Textbook on the Agriculture of the Sugar Cane, the Manufacture of Cane Sugar, and the Analysis of Sugar House Products*. 2nd edition. London: Norman Rodger, 1921.

———. *The History of Sugar*. 2 vols. London: Chapman & Hall, 1949–1950.

De Nettancourt, Gabrielle. "Le peuplement néerlandais à l'île Maurice" (The Dutch peopling of Mauritius). *Actes du colloque sur le mouvement des populations dans l'Océan Indien* (1971): 219–231.

Desjardins, Julien. *Notice historique sur Charles Telfair* (A historical note on Charles Telfair). Port Louis: 1836.

De Sornay, Aimé. "La canne à sucre M.134/32" (The sugar cane M.134/32). *Revue agricole et sucrière de l'Ile Maurice* 32 (Jan.- Feb.1953): 5–11.

———. "Méthode d'obtention de nouvelles variétés de cannes" (Method of obtaining new cane varieties). *Revue agricole et sucrière de l'Ile Maurice* 51 (May-June 1930): 99–103.

———. "Resistance of Sugar Cane Varieties to Cyclones," *Revue agricole et sucrière de l'île Maurice* 37, nos.4–5 (1958): 241–245.

De Sornay, Pierre. "Historique de la canne de graine à l'île Maurice" (An account of seedling canes in Mauritius). *Revue agricole et sucrière de l'île Maurice* 10, no.58 (1931): 125–126.

Domaingue, Robert. "Canne à sucre ou canne à fibre" (Sugar-cane or fiber-cane). *PROSI: Agriculture, Commerce, Industrie*, December 1990, pp.6–7.

Drayton, Richard Harry. "Imperial Science and a Scientific Empire: Kew Gardens and the Uses of Nature, 1772–1903." Ph.D. diss., Department of History, Yale University, 1993.

———. "Science and the European Empires," *Journal of Imperial and Commonwealth History* 23:3 (1996): 503–10.

———. "Sugar Cane Breeding in Barbados: Knowledge and Power in a Colonial Context." A.B. thesis, Department of the History of Science, Harvard University, 1986.

Du Petit-Thouars, Aubert. *Mélanges de botanique et de voyages* (Mixtures of botany and travels). Paris: A. Bertrand, 1811.

Dunstan, Wyndham R. "The Work of the Imperial Institute for India." *Bulletin of the Imperial Institute* 14 (1916): 183–227.

Edgerton, Claude W. *Sugarcane and Its Diseases*. Baton Rouge, Louisiana: Louisiana State University Press, 1955.

Eriksen, Thomas Hylland. *Communicating Cultural Difference and Identity*. University of Oslo Occasional Paper in Social Anthropology No.16. Oslo: University of Oslo, 1988.

———. "Creole Culture and Social Change." *Journal of Mauritian Studies* 1, no.2 (1986): 59–72.

———. "Nationalism, Mauritian Style: Cultural Unity and Ethnic Diversity." *Comparative Studies in Society and History* 36, no.3 (July 1994): 549–74.

Evans, H. "A Brief Review of the Work of the Mauritius Sugarcane Research Station, 1930–1948." *Revue agricole et sucrière de l'île Maurice* 28, no.1 (1949): 12–20.

Fairhead, James, and Melissa Leach, *Misreading the African Landscape: Society and Ecology in a Forest-Savanna Mosaic*. Cambridge: Cambridge University Press, 1996.

Farmer, B.H. "Perspectives on the Green Revolution in South Asia." *Modern Asian Studies* 20, no.1 (1986): 175–199.

Feierman, Steven. *Peasant Intellectuals: Anthropology and History in Tanzania*. Madison: University of Wisconsin Press, 1990.

Ffrench-Mullen, M.D. "Agricultural Teaching in Mauritius in Relation to Sugar Cane." *Revue agricole et sucrière de l'île Maurice* 43, no.3 (1964): 316–320.

Filliot, Jean-Michel. *La traite des esclaves vers les Mascareignes au XVIIIe siècle* (The slave trade around the Mascarenes during the eighteenth century). Mémoires ORSTOM No.72. Paris: ORSTOM, 1974.

Fitzgerald, Deborah. *The Business of Breeding: Hybrid Corn in Illinois, 1890–1940*. Ithaca, New York: Cornell University Press, 1990.

Foucault, Michel. *The Order of Things: An Archaeology of the Human Sciences.* New York: Pantheon, 1970.

Frey, Kenneth J., ed. *Historical Perspectives in Plant Science.* Ames: Iowa State University Press, 1994.

Fusée Aublet, Jean Baptiste Christophere. *Histoire des plantes de la Guiane Françoise . . . & une notice des plantes de l'Isle de France* (A history of the plants of French Guiana . . . and a note on the plants of Mauritius). 4 Vols. Paris: Didot, 1775.

Galloway, J.H. *The Sugar Cane Industry: An Historical Geography from Its Origins to 1914.* Cambridge: Cambridge University Press, 1989.

Geison, Gerald L. *The Private Science of Louis Pasteur.* Princeton: Princeton University Press, 1995.

Gerbeau, Hubert. "The Slave Trade in the Indian Ocean: Problems Facing the Historian and Research to Be Undertaken." In *The African Slave Trade from the Fifteenth to the Nineteenth Century.* Paris: UNESCO, 1979, 184–207.

Gilroy, Paul. *The Black Atlantic: Modernity and Double Consciousness.* Cambridge: Harvard University Press, 1993.

Golant, William. *Image of Empire: The Early History of the Imperial Institute, 1887–1925.* Pamphlet. Exeter: University of Exeter Press, 1984.

Govinden, N. "Intercropping of Sugar Cane with Potato in Mauritius: A Successful Cropping System," *Field Crops Research* 25 (1990): 99–110.

Graham, Gerald S. *Great Britain in the Indian Ocean: A Study of Maritime Enterprise, 1810–1850.* Oxford: Clarendon Press, 1967.

Gramsci, Antonio. *Selections from the Prison Notebooks.* Ed. Quintin Hoare and Geoffrey Nowell Smith. New York: International Publishers, 1971.

Grant, Baron Charles. *The History of Mauritius or the Isle de France.* London: Bulmer, 1801.

Green, R.H., and S.H. Hymer. "Cocoa in the Gold Coast : A Study in the Relations between African Farmers and Agricultural Experts." *Journal of Economic History* 26, no.3 (1966): 299–319.

Griffing, Bruce. "Historical Perspectives on Contributions of Quantitative Genetics to Plant Breeding." In *Historical Perspectives in Plant Science.* Ed. Kenneth J. Frey. Ames: Iowa State University Press, 1994, 43–86.

Grove, Richard H. *Green Imperialism: Colonial Expansion, Tropical Island Edens and the Origins of Environmentalism, 1600–1860.* Cambridge: Cambridge University Press, 1995.

Guilding, Lansdown. *An Account of the Botanic Garden in the Island of Saint Vincent from Its First Establishment to the Present Time.* Glasgow: Richard Griffin and Co., 1825; reprint edition, Forge Village, Massachusetts: Murray Printing Co., 1964.

Haas, Peter M. *Saving the Mediterranean: The Politics of International Environmental Cooperation.* New York: Columbia University Press, 1990.

Habermas, Jürgen. *Jürgen Habermas on Society and Politics: A Reader.* Ed. Steven Seidman. Boston: Beacon Press, 1989.

Haddom, E. "Milling of the Uba Cane in South Africa." *Revue agricole et sucrière de l'île Maurice* 8, no.43 (1929): 13–14.

Hall, Douglas. *Planters, Farmers, and Gardeners in Eighteenth-Century Jamaica.* The 1987 Elsa Goveia Memorial Lecture. Mona, Jamaica: Department of History, University of the West Indies, 1987.

Harland, S.C. "Plant Breeding and Genetics." *Empire Cotton Growing Review* 32, no.1 (1955): 19–23.

Hazareesingh, K. *History of Indians in Mauritius.* London: Macmillan, 1975.

Headrick, Daniel. *The Tentacles of Progress: Technology Transfer in the Age of Imperialism, 1850–1940.* Oxford: Oxford University Press, 1988.

———. *The Tools of Empire: Technology and European Imperialism in the Nineteenth Century.* Oxford: Oxford University Press, 1981.

Hess, David J. *Science and Technology in a Multicultural World: The Cultural Politics of Facts and Artifacts.* New York: Columbia University Press, 1995.

Hookoomsing, Vinesh Y., "L'emploi de la langue Créole dans le contexte multilingue et multiculturel de l'île Maurice" (The use of the Creole language in the multilingual and multicultural context of Mauritius). Ph.D. diss., Department of Linguistics, Université Laval, Québec, 1987.

———. "So Near, Yet So Far: *Bannzil's* Pan-Creole Idealism." *International Journal of the Sociology of Language* 102 (1993): 27–38.

Houbert, Jean. "Mauritius: Independence and Dependence." *The Journal of Modern African Studies* 19, no.1 (1981): 75–105.

"The Imperial Institute and the Dominions Royal Commission." *Bulletin of the Imperial Institute* 15 (1917): 184–198.

"Interview with Bruce Campbell." *PROSI: Agriculture, Commerce, Industrie*, September 1988, 22–24.

Jarrell, Richard A. "Differential National Development and Science in the Nineteenth Century: The Problems of Quebec and Ireland." In *Scientific Colonialism: A Cross-Cultural Comparison.* Ed. Nathan Reingold and Marc Rothenberg. Washington, D.C.: Smithsonian Institution Press, 1987, 323–350.

Jeffries, Charles. "Sir Frank Arthur Stockdale," *Dictionary of National Biography* (1941–1950): 840–841.

Jeswiet, Jakob. "The History of Sugar-Cane Selection Work in Java." *Proceedings of the International Society of Sugar Cane Technologists* (1927): 115–122.

Kennedy, Paul M. *Preparing for the Twenty-First Century.* New York: Random House, 1993.

Kloppenburg, Jack R. *First the Seed: The Political Economy of Plant Biotechnology, 1492–2000.* Cambridge: Cambridge University Press, 1988.

Knight, Franklin W. *The Caribbean: The Genesis of a Fragmented Nationalism.* Second Edition. Oxford: Oxford University Press, 1990.

Koenig, Jacques. "The Commonwealth Sugar Agreement." In *The Sugar Protocol and Mauritius.* Ed. Bernard Boullé. Port Louis: Editions IPC, 1989, 1–26.

———. *Mauritius and Sugar.* Les Pailles, Mauritius: Précigraph, 1988.

Kohler, Robert E. *Lords of the Fly: Drosophila Genetics and the Experimental Life.* Chicago: University of Chicago Press, 1994.

Kuczynski, R.R. *Demographic Survey of the British Colonial Empire.* 2 Vols. Oxford: Oxford University Press, 1949.

Lagesse, Marcelle. *L'île de France avant La Bourdonnais* (Mauritius before Labourdonnais). Mauritius Archives Publication No. 12. Port Louis: Imprimerie Commerciale, 1972.

Laitin, David D. *Hegemony and Culture: Politics and Religious Change among the Yoruba.* Chicago: University of Chicago Press, 1986.

Lamusse, Roland. "The Economic Development of the Mauritius Sugar Industry: Mauritius Sugar in World Trade." *Revue agricole et sucrière de l'île Maurice* 44, no.1 (1965): 11–36.

———. "The Economic Development of the Mauritius Sugar Industry: Labour Problems." *Revue agricole et sucrière de l'île Maurice* 43, no.2 (1964): 113–127.

———. "The Economic Development of the Mauritius Sugar Industry: Development in Field and Factory." *Revue agricole et sucrière de l'île Maurice* 43, no.1 (1964): 22–38.

———. "The Economic Development of the Mauritius Sugar Industry: The Sources of Capital and System of Crop Finance." *Revue agricole et sucrière de l'Ile Maurice* 43, no.4 (1964): 354–372.

Latour, Bruno. *The Pasteurization of France.* Cambridge: Harvard University Press, 1988.

———. *Science in Action: How to Follow Scientists and Engineers through Society.* Cambridge: Harvard University Press, 1987.

Law, John. "Technology and Heterogeneous Engineering: The Case of Portuguese Expansion." *The Social Construction of Technological Systems: New Directions in the Sociology and History of Technology.* Ed. Wiebe E. Bijker et al. Cambridge: MIT Press, 1987, 111–134.

Lehembre, Bernard. *L'île Maurice* (Mauritius). Paris: Karthala, 1981.

Leonard, David K. *Reaching the Peasant Farmer: Organization Theory and Practice in Kenya.* Chicago: University of Chicago Press, 1977.

Lincoln, H. *The Past and Present Position of the Mauritius Sugar Industry.* Port Louis: General Printing and Stationery, 1938.

Lipton, Michael, with Richard Longhurst. *New Seeds and Poor People*. Baltimore: The Johns Hopkins University Press, 1989.

Lutchmeenaraidoo, K. "Agricultural Extension: New Concepts and Limitations." *Revue agricole et sucrière de l'île Maurice* 48, no.3 (1969): 190–194.

Ly-Tio-Fane Pineo, Huguette. *Lured Away: The Life History of Indian Cane Workers in Mauritius*. Moka, Mauritius: Mahatma Gandhi Institute, 1984.

Ly-Tio-Fane, Madeleine. "Contacts between Schönbrunn and the Jardin du Roi at Isle de France (Mauritius) in the Eighteenth Century: An Episode in the Career of Nicolas Thomas Baudin." *Mitteilungen des Osterreichischen Staatsarchivs* 35 (1982): 85–109.

———. *Mauritius and the Spice Trade: The Odyssey of Pierre Poivre*. Mauritius Archives Publication No.4. Port Louis: Esclapon, 1958.

———. *Notice historique publiée à l'occasion du cent-cinquantenaire de la Société Royale des Arts et des Sciences de l'île Maurice, 1829–1979* (A historical note published on the occasion of the 150th anniversary of the Royal Society of Arts and Sciences of Mauritius). Les Pailles, Mauritius: Henry & Cie., 1979.

———. "A Reconnaissance of Tropical Resources during Revolutionary Years: The Role of the Paris Museum d'Histoire Naturelle." *Archives of Natural History* 18, no.3 (1991): 333–362.

———. *The Triumph of Jean-Nicolas Céré and His Isle Bourbon Collaborators*. Paris: Mouton, 1970.

McClellan, James E. *Colonialism and Science: Saint Domingue in the Old Regime*. Baltimore: The Johns Hopkins University Press, 1992.

MacDougall, Walter A. *The Heavens and the Earth: a Political History of the Space Age*. New York: Basic Books, 1985.

McIntyre, Guy. "Evolution of Cultural Operations in the Mauritian Sugar Industry." *Revue agricole et sucrière de l'île Maurice* 63, no.2 (1984): 55–61.

MacLeod, Roy. "On Visiting the 'Moving Metropolis': Reflections on the Architecture of Imperial Science." In *Scientific Colonialism: A Cross-Cultural Comparison*. Ed. Nathan Reingold and Marc Rothenberg. Washington, D.C.: Smithsonian Institution Press, 1987, 217–245.

———. "Passages in Imperial Science: From Empire to Commonwealth." *Journal of World History* 4, no.1 (1993): 117–150.

Maier, Charles S. *Recasting Bourgeois Europe: Stabilization in France, Germany, and Italy in the Decade after World War I*. Princeton: Princeton University Press, 1975.

Malim, Michael. *Island of the Swan*. London: Longman, 1953.

Malleret, Louis. *Pierre Poivre*. Paris: Ecole Française d'Extrême Orient, 1974.

Manrakhan, Jagadish. *The Mauritian School for Scientific Agriculture, 1914 -1989*. Le Réduit, Mauritius: University of Mauritius, 1991.

———. "The Role of Marketing in the Agricultural Development of Mauritius." *Revue agricole et sucrière de l'île Maurice* 48, no.3 (1969): 195–201.

Marx, Jean L., ed. *A Revolution in Biotechnology*. Cambridge: Cambridge University Press, 1989.

Masefield, G.B. *A History of the Colonial Agricultural Service*. Oxford: Clarendon Press, 1972.

Mathur, Raj. "The Battle for a Partly-Elective Legislature." *Journal of Mauritian Studies* 1, no.2 (1986): 73–108.

The Mauritius Chamber of Agriculture. *The Mauritius Chamber of Agriculture, 1853–1953*. Port Louis: General Printing and Stationery Co., 1953.

———. *Report on Planters' Sugar Cane Yields*. Port Louis: 1971.

———. *Report of the Reviewing Committee on Planters' Yields*. Port Louis: June 1975.

———. *Reviewing Committee on Planters' Sugar Cane Yields: Second Report*. Port Louis, December 1980.

Meade, J.E., et al. *The Economic and Social Structure of Mauritius*. London: Frank Cass, 1961.

Meagher, Kate. "Institutionalizing the Bio-Revolution: Implications for Nigerian Smallholders." *Journal of Peasant Studies* 18, no.1 (1990): 68–89.

Moore, P.H., and J.E. Irvine. "Genomic Mapping of Sugarcane and Its Potential Contribution to Improvement and to Selection of New Varieties." *Proceedings of the South African Sugar Technologists' Association* (June 1991): 96–102.

Morgan, David John. *The Official History of Colonial Development*. Vol.1. *The Origins of British Aid Policy, 1924–1945*. London: Macmillan, 1980.

Morris, Daniel, and Frank Stockdale, "The Improvement of the Sugar-Cane by Selection and Hybridization." *West Indian Bulletin* 7 (1906): 345–73.

North-Coombes, Alfred. "Aperçu de l'évolution de l'agriculture à l'île Maurice" (A survey of the evolution of agriculture in Mauritius). *Revue agricole et sucrière de l'île Maurice* 42, no.4 (1963): 225–233.

———. *The Evolution of Sugarcane Culture in Mauritius*. Port Louis: 1937.

———. *La découverte des Mascareignes par les Arabes et les Portugais* (The discovery of the Mascarenes by the Arabs and the Portuguese). Port Louis: Henry & Cie., 1979.

———. *A History of Sugar Production in Mauritius; Being a More Appropriate Title than "The Evolution of Sugarcane Culture in Mauritius" as Originally Published in 1937, and with a New Opening Chapter*. Port Louis: Mauritius Printing Specialists, 1993.

———. "The Royal Botanical Gardens, Pamplemousses," *Revue agricole et sucrière de l'île Maurice* 42, no.1 (1963): 36–41.

———. *The Vindication of François Leguat*. 2nd edition. Rose Hill, Mauritius: Editions de l'Océan Indien, 1991.

North-Coombes, F. *Mes champs et mon moulin* (My fields and my mill). Port Louis: 1950.

North-Coombes, M.D. "From Slavery to Indenture: Forced Labour in the Political Economy of Mauritius, 1834–1867." In *Indentured Labour in the British Empire, 1834–1920.* Ed. Kay Saunders. London: Croom Helm, 1984.

———. "Struggles in the Cane Fields: Small Cane Growers, Millers, and the Colonial State in Mauritius, 1921–1937." *Journal of Mauritian Studies* 2, no.1 (1987): 1–44.

Northrup, David. *Indentured Labor in the Age of Imperialism, 1834–1922.* Cambridge: Cambridge University Press, 1995.

Nwulia, Moses D.E. *The History of Slavery in Mauritius and the Seychelles, 1810–1875.* East Brunswick, New Jersey: Associated University Presses, 1981.

Orsenigo, Luigi. *The Emergence of Biotechnology: Institutions and Markets in Industrial Innovation.* New York: St. Martin's Press, 1989.

Osborne, Michael A. *Nature, the Exotic, and the Science of French Colonialism.* Bloomington: Indiana University Press, 1994.

Padya, B.M. *Weather and Climate of Mauritius.* Moka: Mahatma Gandhi Institute Press, 1989.

Pain, Adam. "Agricultural Research in Sri Lanka: An Historical Account." *Modern Asian Studies* 20, no.4 (1986): 755–778.

Palladino, Paolo. "Science, Technology, and the Economy: Plant Breeding in Great Britain, 1920–1970." *Economic History Review* 49, no.1 (1996): 116–136.

Palladino, Paolo, and Michael Worboys. "Science and Imperialism," *Isis* 84, no.1 (March 1993): 91–102.

Parris, G.K. "James W. Parris: Discoverer of Sugarcane Seedlings." *The Garden Journal* 4, no.5 (1954): 144–146.

Paturau, J. Maurice. *Histoire économique de l'île Maurice* (Economic history of Mauritius). Les Pailles, Mauritius: Henry & Cie., 1988.

Pike, Nicolas. *Sub-Tropical Rambles in the Land of the Aphanapteryx: Personal Experiences, Adventures, and Wanderings in and around Mauritius.* New York: Harper, 1873.

Pillay, Kessawa P. Payandi. "Redesigning the Extension Services in Mauritius to Meet the Needs of the Small Sugarcane Planters." M.Sc. thesis, University of Reading, 1989.

Pitot, Albert. *T'Eylandt Mauritius: esquisses historiques* (The island of Mauritius: historical sketches). Port Louis: Coignet, 1905.

Porter, Theodore M. *The Rise of Statistical Thinking, 1820–1900.* Princeton: Princeton University Press, 1986.

Pridham, Charles. *England's Colonial Empire: An Historical, Political and Statistical Account of the Empire, Its Colonies, and Dependencies.* Vol. 1. *The Mauritius and Its Dependencies.* London: Smith, Elder & Co., 1846.

Proctor, Robert N. *Value-Free Science? Purity and Power in Modern Knowledge.* Cambridge: Harvard University Press, 1991.

Public Relations Office of the Sugar Industry (PROSI). *L'industrie sucrière de l'île*

Maurice (The sugar industry of Mauritius). Fourth revised edition. Les Pailles, Mauritius: Henry & Cie., 1989.

Pyenson, Lewis. *Civilizing Mission: Exact Sciences and French Overseas Expansion, 1830–1940*. Baltimore: The Johns Hopkins University Press, 1993.

Ram, Sooresh. "MSIRI: 40 ans de recherche" (MSIRI: Forty years of research). *Business Magazine* 24 (25 September 1992): 21–29.

Ramdin, T. *Mauritius: A Geographical Survey*. London: University Tutorial Press, 1969.

Reddi, Sadasivam J. "Aspects of Slavery during the British Administration." In *Slavery in South-West Indian Ocean*. Ed. U. Bissoondoyal. Moka, Mauritius: Mahatma Gandhi Institute, 1989, 106–123.

———. "The Development of Political Awareness among Indians, 1870–1930." *Journal of Mauritian Studies* 3, no.1 (1989): 1–15.

———. "The Establishment of the Indian Indenture System, 1834–1842." In *Indians Overseas: the Mauritian experience*. Ed. U. Bissoondoyal. Moka, Mauritius: Mahatma Gandhi Institute, 1984, 1–17.

———. "Labour Protest among Indian Immigrants." In *Indians Overseas: The Mauritian Experience*. Ed. U. Bissondoyal. Moka, Mauritius: Mahatma Gandhi Institute, 1984, 277–300.

Revue agricole et sucrière de l'île Maurice. 1922–Present.

Ricaud, C., et al. *Diseases of Sugarcane: Major Diseases*. Amsterdam: Elsevier, 1989.

Richards, Paul. *Indigenous Agricultural Revolution: Ecology and Food Production in West Africa*. London: Hutchinson, 1985.

Robinson, Ronald. "Non-European Foundations of European Imperialism: Sketch for a Theory of Collaboration." In *Studies in the The Theory of Imperialism*. Eds. Roger Owen and Bob Sutcliffe. London: Longman, 1972.

Rossiter, Margaret W. *The Emergence of Agricultural Science: Justus Liebig and the Americans, 1840–1880*. New Haven: Yale University Press, 1975.

Rouillard, Guy. *Histoire des domaines sucriers de l'île Maurice* (The history of the sugar estates of Mauritius). Les Pailles, Mauritius: Henry & Cie., 1964–1979.

———. *Historique de la canne à sucre à l'île Maurice, 1639–1989* (A history of sugar cane in Mauritius). Port Louis: MSM, 1990.

Rouillard, Guy, with Joseph Guého. *Le jardin des Pamplemousses, 1729–1979* (The Pamplemousses garden). Les Pailles, Mauritius: Henry & Cie., 1983.

Rowe, William T. "The Public Sphere in Modern China." *Modern China* 16, no.3 (1990): 309–329.

Roy, Jay Narain. *Mauritius in Transition*. Allahabad: Tripathi, 1960.

Ryckebusch, Jacky. *Bertrand-François Mahé de la Bourdonnais: entre les Indes et les Mascareignes* (Between the Indies and the Mascarenes). Sainte-Clotilde, Réunion: Centre de Recherche Indianocéanique, 1989.

Schama, Simon. *Landscape and Memory*. New York: Alfred A. Knopf, 1995.

Scott, Joan W. *The Glassworkers of Carmaux: French Craftsmen and Political Action in a Nineteenth-Century City.* Cambridge: Harvard University Press, 1974.

Shapin, Steven. *The Scientific Revolution.* Chicago: University of Chicago Press, 1996.

Shapin, Steven, and Simon Schaffer, *Leviathan and the Air-Pump: Hobbes, Boyle, and the Experimental Life.* Princeton: Princeton University Press, 1985.

Shiva, Vandana. *The Violence of the Green Revolution: Third World Agriculture, Ecology, and Politics.* London: Zed Books, 1991.

Simmonds, N. W., Ed. *Evolution of Crop Plants.* London: Longman, 1976.

Simmons, Adele Smith. *Modern Mauritius: the Politics of Decolonization.* Bloomington: Indiana University Press, 1982.

So, Alvin Y. *Social Change and Development: Modernization, Dependency, and World-System Theory.* Newbury Park: Sage Publications, 1990.

Starr, Cecie, and Ralph Taggart. *Biology: The Unity and Diversity of Life.* Fifth Edition. Belmont, California: Wadsworth Publishing Inc., 1989.

Stebbins, G. Ledyard. "Biological Revolutions of Thought during the Twentieth Century." In *Historical Perspectives in Plant Science.* Ed. Kenneth J. Frey. Ames: Iowa State University Press, 1994, 3–21.

Stevenson, G.C. *Genetics and Breeding of Sugar Cane.* London: Longman, 1965.

———. "Sugar Cane Varieties in Barbados: An Historical Overview." *British West Indies Central Sugar Cane Breeding Station Bulletin* 37 (1954).

———. "Sugar-Cane Varieties in British Guiana: An Historical Review." *Empire Journal of Experimental Agriculture* 16, no.63 (1948): 143–154.

Storey, William Kelleher. "Biotechnology and Power: Farmers, the Colonial State, and the Quest for Better Sugar Cane in Mauritius, 1853–1953." Ph.D. diss., Department of History, The Johns Hopkins University, 1993.

Summers, Carol. *From Civilization to Segregation: Social Ideas and Social Control in Southern Rhodesia, 1890–1934.* Athens: Ohio University Press, 1994.

Swettenham, Frank. *Also and Perhaps.* London: Bodley Head, 1912.

Telfair, Charles. *Some Account of the State of Slavery at Mauritius, Since the British Occupation in 1810, in Refutation of Anonymous Charges Promulgated against Government and that Colony* Port Louis: 1830

Tinker, Hugh. *A New System of Slavery: The Export of Indian Labour Overseas, 1830–1920.* London: Oxford University Press, 1974.

Tempany, H.A., and D.H. Grist. *An Introduction to Tropical Agriculture.* London: Longman, 1958.

Titmuss, Richard M., and Brian Abel-Smith. *Social Policies and Population Growth in Mauritius.* London: Frank Cass and Co., 1961.

Todd, Jan. *Colonial Technology: Science and the Transfer of Innovation to Australia.* Cambridge: Cambridge University Press, 1995.

Toussaint, Auguste. *Early Printing in the Mascarene Islands, 1767–1810.* London: University of London Press, 1951.

————. *Histoire des îles Mascareignes* (History of the Mascarene Islands). Paris: Berger-Levrault, 1972.

————. *Historique de la Chambre d'Agriculture, île Maurice, suivi d'une analyse des rapports annuels, 1854–1951* (An account of the Mauritius Chamber of Agriculture, following an analysis of the annual reports). Mimeograph in the library of the Mauritius Chamber of Agriculture. Port Louis: Mauritius Chamber of Agriculture, 1953.

————. *History of Mauritius.* Trans. W.E.F. Ward. London: Macmillan, 1977.

————. *History of the Indian Ocean.* Trans. June Guicharnaud. London: Routledge & Kegan Paul, 1966.

————. *Le mirage des îles: le négoce français aux Mascareignes au XVIIIème siècle* (The mirage of islands: French trade in the Mascarenes during the eighteenth century). Aix-en-Provence: Edisud, 1977.

————. *La route des îles: contribution à l'histoire maritime des Mascareignes* (The course of the islands: a contribution to the maritime history of the Mascarenes). Paris: SEVPEN, 1967.

Toussaint, Auguste, and Harold Adolphe. *Bibliography of Mauritius, 1502–1954.* Port Louis: Esclapon, 1956.

Toussaint, Auguste, et al., comps. *Dictionary of Mauritian Biography.* Société de l'Histoire de l'Ile Maurice. Port Louis: 1944–1975.

Twain, Mark. *More Tramps Abroad.* London: Chatto & Windus, 1897.

Tyack, Jean-Claude. "Pensées vertes: l'environnement et l'industrie sucrière" (Green thoughts: the environment and the sugar industry). *PROSI: Agriculture, Commerce, Industrie,* June 1992, 13–14.

U.S. Department of Agriculture. Agricultural Research Service. "In Search of Perfect Plants: ARS Plant Breeding Research Today." *Agricultural Research: Bringing Forth the Best* 38, no.10 (1990): 3–10.

Van den Ban, A., and H.S. Hawkins. *Agricultural Extension.* New York: John Wiley, 1988.

Venkatraman, Rao Bahadur T. S. "Methods of Selecting Seedlings as Adopted at Coimbatore." *Proceedings of the International Society of Sugar Cane Technologists* (1935): 344–348.

————. "Problems for the Sugar Cane Breeder with Special Reference to Indian Conditions." *Proceedings of the International Society of Sugar Cane Technologists* (1929): 429–449.

Verin, Pierre. *Maurice avant l'Isle de France* (Mauritius before the Isle de France). Paris: Fernand Nathan, 1983.

Wallerstein, Immanuel. *The Modern World-System.* 3 Vols. New York: Academic Press, 1974, 1980, 1989.

Wanquet, Claude. *Histoire d'une révolution: la Réunion, 1789–1803* (History of a revolution: Reunion Island). Marseille: Jeanne Lafitte, 1980.

———. "Le café à la Réunion, une civilisation disparue" (Coffee on Reunion Island, a vanished civilization). In *Fragments pour une histoire des économies et sociétés de plantation*. Ed. Claude Wanquet. St. Denis, Réunion: Université de la Réunion, 1989, 55–73.

———. "Les fondements historiques de la cooperation régionale" (The historical foundations of regional cooperation). *Annuaire des pays de l'Océan Indien* 9 (1982–1983): 21–45.

Wiehe, J.B., N. Govinden, and P. Rouillard. "Achievements and Prospects in Crop Diversification on Sugar Cane Lands in Mauritius." *Revue agricole et sucrière de l'île Maurice* 63, no.2 (1984): 135–147.

Willemet, Pierre Rémi François de Paule. *Herbarium Mauritianum*. Leipzig: P.P. Wolf, 1796.

Worboys, Michael. "Science and the Colonial Empire, 1895–1940." In *Science and Empire: Essays in Indian Context, 1700–1947*. Ed. Deepak Kumar. Delhi: n.p., 1991, 13–27.

The World Bank. *Mauritius: Economic Memorandum: Recent Developments and Prospects*. A World Bank Country Study. Washington, D.C.: 1983.

———. *Mauritius: Expanding Horizons*. Report No. 9685–MAS. Washington, D.C.: June 12, 1992.

———. *Mauritius: Managing Success*. A World Bank Country Study. Washington, D.C.: 1989.

———. *Staff Appraisal Report: Mauritius: Agricultural Management and Services Project*. Report No. 9330–MAS. Washington, D.C.: May 7, 1991.

———. *Staff Appraisal Report: Mauritius: Sugar Industry Project*. Report No.6092–MAS. Washington, D.C.: June 5, 1986.

Yanchinski, Stephanie, ed. *Biotechnology: A Brave New World?* Cambridge: Lutterworth Press, 1989.

Yudelman, Montague. "The Transfer of Agricultural Techniques." In *Colonialism in Africa, 1870–1960* Vol. 4. *The Economics of Colonialism*. Ed. Peter Duignan and L.H. Gann. Cambridge: Cambridge University Press, 1969, 329–359.

INDEX

Lightning Source UK Ltd.
Milton Keynes UK
UKHW040742160420
361758UK00002B/25/J

9 781580 460156